Andreas Wittwer

Tectonic framework of the central Java subduction zone

Andreas Wittwer

Tectonic framework of the central Java subduction zone

This work is based on data of the joint interdisciplinary MERAMEX (Merapi Amphibious Experiment) project

Südwestdeutscher Verlag für Hochschulschriften

Imprint
Any brand names and product names mentioned in this book are subject to trademark, brand or patent protection and are trademarks or registered trademarks of their respective holders. The use of brand names, product names, common names, trade names, product descriptions etc. even without a particular marking in this work is in no way to be construed to mean that such names may be regarded as unrestricted in respect of trademark and brand protection legislation and could thus be used by anyone.

Publisher:
Südwestdeutscher Verlag für Hochschulschriften
is a trademark of
Dodo Books Indian Ocean Ltd., member of the OmniScriptum S.R.L Publishing group
str. A.Russo 15, of. 61, Chisinau-2068, Republic of Moldova Europe
Printed at: see last page
ISBN: 978-3-8381-2637-1

Zugl. / Approved by: Kiel, Christian-Albrechts-Universität, Diss., 2011

Copyright © Andreas Wittwer
Copyright © 2011 Dodo Books Indian Ocean Ltd., member of the OmniScriptum S.R.L Publishing group

Contents

1 Introduction   1
- 1.1 Sunda Arc: Geodynamic setting 4
- 1.2 Seismicity 7
- 1.3 GINCO (1999):
  Investigations along southern Sumatra to western Java 9
- 1.4 SINDBAD (2006):
  Investigations along eastern Java to Bali and Lombok 11
- 1.5 MERAMEX project (2004) and motivation (this study) 12

2 Seismic data   15
- 2.1 Data acquisition 15
- 2.2 Data processing 16

3 Modeling the seismic P-wave velocities from wide-angle data   19
- 3.1 Forward modeling with raytracing 20
- 3.2 Inverse modeling with 2D tomography 21
  - 3.2.1 Forward problem 22
    - 3.2.1.1 Shortest path method (SPM) 22
    - 3.2.1.2 Ray bending 23
  - 3.2.2 Inverse problem 24
  - 3.2.3 Inversion parameterization 27
- 3.3 Results of the forward and the inverse modeling 31
  - 3.3.1 Testing different input models for the tomography 31
  - 3.3.2 Final forward and inverse P-wave velocity models 32
  - 3.3.3 Ocean basin, trench and subducted plate 37
    - 3.3.3.1 Obtaining Moho depth by joint inversion of refracted and reflected traveltimes 38
  - 3.3.4 Outer- and inner wedge, backstop 46
  - 3.3.5 Forearc basin, margin wedge 52

		3.3.6	Shelf area	62
		3.3.7	Model uncertainties and sensitivity tests	72
	3.4	Conclusion		78
4	Gravity			79
5	Discussion			85
	5.1	Oceanic lithosphere and the Christmas Island seamount province		85
	5.2	Interplate processes		91
		5.2.1	Segmentation of the forearc high	91
		5.2.2	Critical taper analysis	94
		5.2.3	Seamount subduction and the uplifted forearc high	95
		5.2.4	Forearc basin and submarine landslides	99
	5.3	Seismicity and megathrust earthquake potential		101
	5.4	Comparison of the onshore and offshore tomography results		106
6	Conclusion			109
A	Appendix of inverted wide-angle traveltimes			111

List of Figures

1.1 Two distinct types of subduction zones (modified after von Huene et al. (2009)) are presented in this sketch. A accretionary, B erosive margins. Please refer to a detailed description in the text. 2

1.2 MERAMEX Investigation area. The GINCO transects from 1999 (RV SONNE cruises SO137 and SO138) are located at southern Sumatra to western Java (thin red). The SO179 MERAMEX transects recorded in 2004 are south of central Java. The Merapi volcano (black circle) is positioned in the elongation of the MERAMEX transect. The dashed white line outlines the dimensions of the Roo Rise. The central Java trench retreats in landward direction. Isolated topographic highs along the forearc high (white arrows) indicate a change in the tectonic regime. The most western SINDBAD transect from 2006 (RV SONNE cruise SO190) is situated in eastern Java (thin red). 3

1.3 Collision of India with Eurasia. India (red) has fractured the plate of east Asia into several subplates (violet, yellow, blue), which have been pushed far to the east and southeast as the collision has progressed. The South China Sea opened as Borneo moved away from China. The sequence starts at 60 million years ago. India pushes blocks of the Eurasiatic plate eastward. Continental margin rocks smear along the margin of Southeast Asia, while the northern part of India thrusts beneath Tibet. (Fig. modified from University of Wisconsin - Green Bay). 5

1.4 Tectonic setting. At the Sunda Arc the Indo-Australian oceanic plate subducts beneath the Eurasian plate with a convergence rate of 6.7 cm/a in a 11°N direction. The crustal age increases from 96 Ma off western Java to 135 Ma off eastern Java. The oblique subduction off Sumatra results in the evolution of fault systems: MW-FZ=Mentawai, SU-FZ=Sumatra, and UK-FZ=Ujong-Kulon fault zone. 6

1.5 Epicenter distribution of earthqakes from 1990 to 2007 (NEIC catalogue) with magnitudes ≥ 1 and ≤ 9. The yellow stars indicate earthquakes described in the text. a: 2006 July 17, Mw=7.7; b: 1994 June 02, Mw=7.8; c: 2006 May 26, Mw=6.4 8

1.6 Tectonic segmentation of the western Java forearc high. This figure is based on Kopp et al. (2002) 10

LIST OF FIGURES

1.7 *Station distribution of the MERAMEX experiment. Onshore more than 100 temporary seismological stations were installed around Merapi volcano (black squares). Offshore a temporary seismological OBS/H network was deployed (open circles). The three wide-angle and reflection seismic profiles SO179-P16 to SO179-P19 were covered with 53 OBS/H stations.* 13

2.1 *Wide-angle record section of OBH62 (profile SO179-18) located on the Javanese shelf in very shallow water depth (632 m). The first arrivals can be traced over the complete profile length. All following seismic record sections are displayed with a reduction velocity of 6 km/s.* 16

2.2 *Wide-angle record section of OBH36 (profile SO179-P16). This data example shows typical phases from stations near trench locations on the upper plate. The phases are denoted as following: Pn= mantle phase (oceanic and margin wedge mantle respectively; PmP= Moho reflection, Poc= refraction from oceanic crust, PtocP= reflection from the top of the oceanic plate, Pg forearc= refraction from the forearc high.)* 17

2.3 *Wide-angle record section of OBH30 (profile SO179-P16). This data example shows typical phases from stations on the forearc basin locations of the upper plate. The phases are denoted as following: Pn= mantle phase (oceanic and margin wedge mantle respectively; PcontP= reflection from continental Moho, Psed= shallow sedimentary refractions, Pg margin= refractions from the margin wedge, Pg forearc= refraction from the forearc high.)* 18

3.1 *Modeling strategy for this study. The forward model (1), based on interactive 2D raytracing provides the input model for the tomography (2,inverse step). The forward model is updated with the model differences δV (3). This alternating forward and inverse calculations are repeated until the model differences δV and the misfit between calculated and observed traveltimes is minimized.* 19

3.2 *Forward star used in the SPM to provide a good ray coverage in all search directions.* 23

3.3 *Comparison of ray paths in SPM method and after refining the path in ray bending method.* 23

3.4 *Inversion parameter test with profile SO179-P16. a) All data sets with three velocity smoothing factors (150 - 250) showing the dependency between the varying horizontal correlation length Lh (at top and bottom of the model), the RMS value, χ^2 and the model roughness. b) After fixing horizontal correlation length with Lh 3 and 6 km, the vertical correlation length was fixed at 0.5 and 2 km. c) A velocity smoothing of 250 satisfies a traveltime variance and a low model roughness.* 28

LIST OF FIGURES

3.5 *Inverted model with horizontal correlation lengths of 2/4 km at the top/bottom of the model, vertical correlation lengths of 0.1/1 km and a velocity smoothing of 200. Smaller correlation lengths results in a loss of model smoothness. The traveltime misfit is low but the models are overfitted.* 29

3.6 *Two different input models to the tomography testing the hypothesis for a subducted seamount. The left panel provide the input (forward-) model with increased (top) and decreased (bottom) velocities, pointed by the arrow. The inverted model (middle panel) and the model difference (right panel) display evidences for higher velocities in the forearc high, close to the backstop.* 31

3.7 *Testing lower mantle velocities on profile SO179-P16. The forward model provides reduced lower velocities in lower crust with 7.2 km/s and in the mantle with 7.7 km/s. The inverted model applies higher velocities to the mantle (8 km/s) and to the lower crust (7.4 km/s). The model difference (right panel) demonstrates the increased velocities (positive isolines) after the inversion by red colors.* 32

3.8 *Forward models of profiles SO179-P18 (top) and SO179-P16 (bottom). The P-wave velocities are color coded and additionally marked by numbers [km/s]. Solid lines indicate layer interfaces. Station locations are indicated by triangles. The grey shaded area is not well resolved. The models are aligned to the trench (dashed line).* 34

3.9 *Final inverted model of profile SO179-P18. The upper left figure displays the input model for the inversion, which corresponds to the forward model based on MacRay. The main tectonic features are labeled. The upper right presents the inversion result based on the Tomo 2D code by (Korenaga et al., 2000) after four iterations. The white patches are poorly resolved regions with less than five rays per cell. The lower left figure displays the difference between the input forward and resulting inverted model. The maximum velocity variation iso-lines are ±0.2 km/s. The lower right figure shows the color coded DWS matrix, which reflects the ray density and the ray path quality.* 35

3.10 *Final inverted model of profile SO179-P16. For a detailed description please refer to the previous Fig. 3.9* 36

3.11 *Water migrated seismic record section of profile SO179-P16 (top) and interpretive line drawing (bottom) superimposed with a velocity-depth model of the inversion. Velocity values are noted as numbers. This section only shows the outer rise. Station locations are indicated by triangles.* 37

3.12 *Top: Moho reflectors for the joint inversion. All models were inverted with the same starting model and same parameters. The calculated Moho reflectors converge in a depth range of 15 ± 0.5 km. Bottom: ray coverage of all reflected PmP phases. The area between stations 38 and 37 is not resolved and the Moho is interpolated.* 38

LIST OF FIGURES v

3.13 *Critical wide-angle reflections to constrain the Moho depth with stronger amplitudes at offsets greater than the critical distance Xc. Red arrows indicate strong amplitudes of PmP reflections.* 39

3.14 *Forward result of OBS42: Wide-angle seismic record section of station OBS42 (profile SO179-P18). Calculated traveltimes are in yellow (middle) with corresponding raypaths (bottom).* 42

3.15 *Inverted model of OBS42: Wide-angle seismic record sections of profile SO179-P18 (reduced with 6 km/s). Top: interpreted seismic phases, Center: Calculated (red) and picked traveltimes (blue) with error bars, Bottom: Corresponding ray paths through the model for profile SO179-P18.* 43

3.16 *Forward model of OBS41, profile SO179-P16. Please refer to Fig. 3.14 for details* . . 44

3.17 *Inverted model of OBS41:Wide-angle seismic record sections of profile SO179-P16 (reduced with 6 km/s). Top: interpreted seismic phases, Center: Calculated (red) and picked traveltimes (blue) with error bars, Bottom: Corresponding ray paths through the model for profile SO179-P16.* 45

3.18 *Water migrated seismic record section of profile SO179-P16 (top) and interpretive line drawing (bottom). This section is the landward continuation of Figure 3.11. Station locations are indicated by triangles. The tomographic velocity-depth model is superimposed and the seismic P-wave velocities [km/s] are notified as numbers.* 47

3.19 *Forward model of OBH47, profile SO179-P18.* 48

3.20 *Inverted model of OBH47.* 49

3.21 *Forward model of OBH36, profile SO179-P16.* 50

3.22 *Inverted Model of OBH36.* 51

3.23 *Forward model of OBH53, profile SO179-P18.* 54

3.24 *Inverted model of OBH53.* 55

3.25 *Forward model of OBH58, profile SO179-P18.* 56

3.26 *Inverted model of OBH58* 57

3.27 *Forward model OBH30, profile SO179-P16.* 58

3.28 *Inverted model of OBH30.* 59

3.29 *Forward model of OBH28, profile SO179-P16.* 60

3.30 *Inverted model of OBH28.* 61

3.31 *Forward model of OBH62, profile SO179-P18.* 64

3.32 Inverted model of OBH62 65

LIST OF FIGURES

3.33 *Water migrated seismic record section of profile SO179-P16 (top) and interpretive line drawing (bottom). Station locations are indicated by triangles. The intersection of dip profiles SO179-P18 and SO179-P16 are marked by dashed lines. The tomographic velocity-depth model is superimposed and the seismic P-wave velocities [km/s] are notified as numbers.* 66

3.34 *Forward models of profiles SO179-P18 (top) and SO179-P16 (bottom). The P-wave velocities are color coded and additionally marked by numbers [km/s]. Station locations are indicated by triangles.* 66

3.35 *Velocity model updates for the final model of profile SO179-P19 are less than < 1.9% after the first iteration. Therefore a velocity damping was not applied. After five iterations model updates became < 0.1%. The RMS misfit of all refracted rays after last iteration is 56 ms. The final χ^2 value is 1.23* 67

3.36 *Forward model OBH66, profile SO179-P19.* 68

3.37 *Inverted model of OBH66* 69

3.38 *Forward model OBH73, profile SO179-P19.* 70

3.39 *Inverted model of OBH73* 71

3.40 *Distribution of traveltime residuals $\Delta t = t_{obs} - t_{calc}$ values after first (left) and last iteration (right). First iteration corresponds to the forward model. Only refracted phases to determine the velocity model* 73

3.41 *Checkerboard test for all three profiles. An alternating pattern of 45x8 km rectangles with values of ± 0.55 km/s positive and negative velocity perturbations was applied to the final models (top). The recovered anomaly pattern after four iterations is displayed below.* 75

3.42 *DWS plot gives an indication of ray coverage and high quality ray paths. Red and yellow ray paths have a high dws value.* 76

3.43 *Alternating positive and negative synthetic anomalies with a maximum amplitude of ± 0.4 km/s. The thin iso-velocity lines in the background are displayed for a better orientation to the model structures.* 76

3.44 *Merged tomographic models in a perspective 3D view. The velocity slices are based on the joint tomographic inversion of reflected and refracted phases. The strike line SO179-P19 includes only refracted phases for the inversion. The velocity fields at the intersections of the independently inverted dip lines to the strike line fit very well.* ... 78

4.1 *Free-air gravity anomalies in the survey area. Ship tracks are in green. This map was compiled by the Federal Institute of Geosciences and Natural Resources (BGR).* 80

LIST OF FIGURES

4.2 Top: Blow up of of Fig. 4.1 with the region of the located large seamount around 12°S and 109°E. Red lines display the NS- and WE trending traces of bathymetry and gravity data below. Bottom, left: NS trending trace of the free air gravity corresponds to the shape of the high resolution bathymetry of the seamount with its large dimensions. Right: WE trending trace with free air gravity and the bathymetry (lower resolution from global ETOPO dataset), with its moat around the seamount. 81

4.3 Calculated free air gravity of profile SO179-P16 (top). Applying a lower mantle density of 3.32 g/cm^3 (grey line) like on the western dip-line, does not match the observed gravity (circles). The calculated gravity fits the observed within a range of ± 10 mGal (residual gravity dashed line). Tectonic units (bottom) are color coded with constant density values. 82

4.4 Calculated free air gravity of profile SO179-P18. 83

4.5 Calculated free air gravity of profile SO179-P19. 83

5.1 Interpretation of the modeling results. The western profile SO179-P18 is displayed at the top, the eastern profile SO179-P16 at the bottom. The solid lines indicate an interpretation of the structural boundaries. The color coded P-wave velocities are displayed in the background with a reduced opacity. The velocity-depth functions of these models (yellow, blue) are compared with a global compilation of v(z) of the Pacific plate (black) by White et al. (1992). 86

5.2 Velocity-depth distribution of all Java profiles. At the top the GINCO profile (Kopp et al., 2001), in the middle the central Java profiles of this study and at the bottom the eastern Java profile, investigated by Shulgin et al. (2009). 88

5.3 A broad band of seamounts incipiently subducts south of central Java. The Christmas Island Seamount Province (CHRISP) is a submarine volcanic province oriented in E-W direction. Contour lines range from 4500 m to 1000 m below the water surface with a contour interval of 1000 m. The Roo Rise is indicated by the gray shaded area. 89

5.4 MERAMEX profiles compared with GINCO and SINDBAD. Qualitative volumes of the active outer wedge. α corresponds to the forearc slope angle, which is measured over $\geq$ 50 km to avoid local anomalies. β is the dip angle of the downgoing plate. The taper angle γ (the sum of α and β) increases from Sumatra to central Java. The forearc high is spearated in different kinematic boundaries. The outer wedge includes the frontal prism (dark blue) and the active Neogene prism (light blue). It is bounded by an out of sequence thrust to the Paleogene inner wegde. 92

5.5 a) This graph outlines the stability field of tapered wedges. The taper of central Java lies in the unstable extensional regime, western Java in the stable wedge (red diamonds). The function of the forearc slope angle α (b) and taper angle γ (c) over the orthogonal convergence rate of the oceanic crust. If the convergence rate exceeds 6,3 cm/a the subduction style is erosive (Clift and Vannucchi, 2004). Black squares indicating accretionary and white circles erosive subduction systems. 94

5.6 High resolution swath mapping of the investigation area. Two squares (A, dashed border and B, solid border) showing the area of perspective view at the right hand side. Station locations marked by red circles. 96

5.7 High resolution swath mapping of the investigation area in a perspective view. A) Displays the forearc basin at the bottom and the forearc high with a view from a landward NE - SW direction. B) View from a seaside SE - NW direction with the oceanic plate, the trench and the forearc high. Station locations marked by red circles. 97

5.8 Seismicity of 500 events during the MERAMEX experiment investigated by the German Research Center for Geosciences (GFZ). The depth distribution is color coded. The yellow stars indicate earthquakes described in the text. a: 2006 July 17, Mw=7.7; b: 1994 June 02, Mw=7.8; c: 2006 May 26, Mw=6.4. 220 earthquake hypocenters are projected in a corridor of 50 km (dashed box) onto the landward extended line SO179-P16 (AB). 101

5.9 This figure is from Wagner et al. (2007). Study area with the onshore experimental setup and the tectonic features of the region. Triangles mark the temporary seismological network. Dots refer to the earthquakes used for the tomography. The red colored area in central Java marks the investigation area covered by passive data, the light blue is covered by active seismic data and the grey area marks both data sets in the uppermost 10 km depth layer. 106

5.10 This figure is from Wagner et al. (2007). P-velocity anomalies of two datasets. Top: only the active dataset was used and is presented in horizontal slices at 5, 10 and 15 km depth. Bottom: Combined dataset of active and passive data. The coastline and the wide-angle seismic profiles are included as black lines. The star indicates the location of the earthquake from 2006 May 26. 107

5.11 Interpretation of the velocity structure in central Java from Wagner et al. (2007). The velocity structure is limited in west to the forearc basin. Black dots display approximately 200 earthquakes, which are located in the region of the tomography. The yellow star indicates the Federal Institute of Geosciences and Natural Resources (BGR) hypocentre of the Java earthquake in 2006 May 26. 108

A.1 Inverted model of OBS39 112
A.2 Inverted model of OBS38 113

A.3	Inverted model of OBH37	114
A.4	Inverted model of OBH35	115
A.5	Inverted model of OBH34	116
A.6	Inverted model of OBH33	117
A.7	Inverted model of OBH32	118
A.8	Inverted model of OBS29	119
A.9	Inverted model of OBH27	120
A.10	Inverted model of OBH43	121
A.11	Inverted model of OBS44	122
A.12	Inverted model of OBS45	123
A.13	Inverted model of OBH46	124
A.14	Inverted model of OBH48	125
A.15	Inverted model of OBH49	126
A.16	Inverted model of OBH51	127
A.17	Inverted model of OBH52	128
A.18	Inverted model of OBH55	129
A.19	Inverted model of OBS56	130
A.20	Inverted model of OBS57	131
A.21	Inverted model of OBH59	132
A.22	Inverted model of OBH60	133
A.23	Inverted model of OBH61	134
A.24	Inverted model of OBH64	135
A.25	Inverted model of OBH65	136
A.26	Inverted model of OBH67	137
A.27	Inverted model of OBH68	138
A.28	Inverted model of OBH69	139
A.29	Inverted model of OBH70	140
A.30	Inverted model of OBH71	141
A.31	Inverted model of OBH72	142
A.32	Inverted model of OBH74	143
A.33	Inverted model of OBH75	144

List of Tables

3.1　Inversion parameters applied to the three wide-angle seismic profiles. . . . .　30
3.2　*Picking uncertainties* . . . .　72
3.3　*Uncertainties of the final P-wave velocity models of the three profiles.* . . . .　74

Abstract

Offshore wide-angle seismic data recorded on ocean bottom instruments of a combined onshore- offshore investigation on the tectonic framework of central Java are presented in this study. The joint interdisciplinary project MERAMEX (Merapi Amphibious Experiment) was carried out to characterize the subduction of the Indo-Australian plate beneath Eurasia. Three marine wide-angle profiles are analyzed by combined forward- and inverse modeling of first and later arrival traveltimes and are integrated together with gravity data. The results of this study are compared with former investigations off southern Sumatra, western Java and eastern Java to obtain a detailed image of the Java margin.

The subduction of the oceanic Roo Rise plateau, located south of central Java, with its thickened and buoyant crust, strongly influences subduction dynamics. The trench is retreated about 60 km in a landward direction. Large scale forearc uplift is manifested in isolated forearc highs, reaching water depths of only 1000 m compared to 2000 m water depth off western Java, and results from oceanic basement relief subduction. The dip angle of the underthrusting oceanic lithosphere is 10° underneath the marine forearc and its crustal thickness increases eastward from 9 - 10 km over a distance of 100 km between both dip profiles off central Java, which is thicker than the global average of 7.4 km. The incipient subduction of a broad band of seamounts off central Java causes frontal erosion of the margin here and leads to mass wasting due to oversteepening of the upper trench wall. The well-developed accretionary wedge off southern Sumatra and western Java diminishes into a small frontal prism with steep slope angles of the upper plate off central Java. This causes a persistent threat for generating tsunamis, which may also be triggered by smaller (Mw 8) earthquakes.

The rough surface of the Indo-Australian plate with its volcanic edifices strongly influences the interplate coupling. A subducted and dismembered seamount is revealed on the eastern profile at the toe of the backstop in 15 km depth. This seamount and similar features present on the megathrust may potentially act as asperities or as barriers to seismic rupture, limiting lateral rupture propagation in the co-seismic phase. Subduction earthquakes with a magnitude ≥ 8 are not observed, while smaller earthquakes frequently occur. A remarkable clustering of earthquakes in the forearc mantle wedge below the shallow forearc Moho may be the seismic expression of seamount detachement.

Zusammenfassung

Im Jahre 2004 wurde das interdisziplinäre Projekt MERAMEX (Merapi Amphibious Experiment) ins Leben gerufen, um den Subduktionsprozess der Indo-Australischen Lithosphärenplatte unter die Eurasische Platte zu untersuchen. Dazu wurden südlich der indonesischen Insel Java drei seismische Weitwinkel-Profile mit Ozean-Boden-Seismometern installiert, um refraktionsseismische Daten zu registrieren. Zusätzlich wurde ein passives, temporäres seismisches Netzwerk mit über 100 Breitbandseismometern um den Vulkan Merapi installiert, um tomographische Studien durchzuführen.

Die Bevölkerung von Java ist ständigen potentiellen Naturgefahren ausgesetzt: der Vulkan Merapi fordert in seinen regelmäSSig wiederkehrenden Aktivitätsphasen zahlreiche Opfer und verursacht immense wirtschaftliche Schäden. Auch die Gefahr von Erdbeben und die oft damit einhergehenden Tsunamis stehen in direktem Zusammenhang mit der Subduktion im Tiefseegraben; so geschehen z.B. am 26. Mai 2006 als ein Beben der Stärke Mw= 6.4 mehr als 5000 Opfer an der Südküste von Java forderte. Ein besseres Verständnis der Subduktionsprozesse ist daher eine Voraussetzung für eine genauere Abschätzung des Gefahrenpotentials entlang der Java-Subduktionszone.

Der Hauptfokus der hier vorgelegten Arbeit liegt auf der Untersuchung der drei marinen Weitwinkelseismischen Profile. Anhand einer Kombination aus Vorwärtsmodellierung und Inversion wurde aus den refraktierten Ersteinsätzen und späteren Reflexionseinsätzen ein Geschwindigkeits-Tiefen-Modell berechnet. Als Zusatzinformation dienten Schweredaten, die in einem Vorwärtsmodell der Dichte modelliert wurden. Die Dichte ist abhängig von der Kompressionswellengeschwindigkeit und lässt sich als zusätzliche Randbedingung für die Geschwindigkeitsmodelle heranziehen. Die Ergebnisse dieser Studie werden mit früheren Arbeiten vor Süd-Sumatra, West-Java und Ost-Java vergleichend diskutiert.

Das ozeanische Roo Plateau, südlich von Zentral-Java, beeinflusst die Subduktion nachhaltig. Die abtauchende ozeanische Kruste ist gegenüber dem globalen Durchschnitt 2 - 3 km mächtiger. Der Tiefseegraben ist um ca. 60 km landeinwärts verschoben und das Vorderbogenhoch ist durch die Subduktion des Roo Plateau gehoben. Die südlich von Java in einem breiten Band auf dem Meeresboden auftretenden vulkanischen Kuppen der Christmas-Island Seamount Province werden in die Subduktionszone getragen und hinterlassen Eintrittsnarben an der Subduktionsfront. Die Modellierungen des östlichen Profils geben deutliche Hinweise auf einen subduzierten Vulkan in 15 km Tiefe, der vermutlich zerschert wurde. Zusammen mit der Subduktion des Roo Plateau führt die Subduktion ausgeprägter Meeresbodentopographie zu einer negativen Massenbilanz und somit zu einer erosiven Subduktion. Die rezent erosive Subduktion vor Zentral-Java führt aktuell zu kaum einer Anlagerung von Sedimenten. Das gesamte östliche Profil ist durch das Roo Plateau kompressiven Kräften ausgesetzt und sowohl an der Subduktionsfront als auch zum Vorderbogen-Sedimentbecken übersteilt. Dies führt zu submarinen Hangrutschungen mit einer wachsenden Gefahr dadurch ausgelöster Tsunamis.

Der Übergang von einer akkretionären zu einer erosiven Subduktion überdeckt lediglich einen Bereich von etwa 100 km zwischen den beiden, entlang der Subduktionsrichtung verlaufenden, Weitwinkelseismischen Profile.

Subduzierte vulkanische Kuppen limitieren während eines Erdbebens potentiell die GröSSe der Bruchzone, die proportional zur Bebenstärke ist. Das erklärt in Zusammenhang mit der Existenz eines flachen Krusten-Mantel-Übergangs in der Oberplatte möglicherweise auch die Tatsache, dass es vor Zentral-Java keine Beben mit Magnituden ≥ 8 gibt, da die laterale Ausdehnung der seismogenen Zone durch die Tiefenlage der Moho der überschiebenden Platte kontrolliert wird. Die Auswertung der MERAMEX Daten zeigt eine vergleichsweise geringe seismische Aktivität der seismogenen Zone. Das lässt Rückschlüsse auf die entweder sehr starke oder nur sehr schwache Kopplung zwischen der Oberplatte und der abtauchenden Platte zu. Weitere Hinweise auf subduzierende submarine Topographie zeigt auch die Seismizität im Mantel der Oberplatte, die auf eine Abscherung von Tiefseekuppen hindeutet, wie dies auch in der japanischen Subduktionszone beobachtet wird.

Chapter 1

Introduction

The Java margin as part of the Sunda Arc is a tectonically highly active subduction zone with a great potential for natural hazards such as volcanism, earthquakes and tsunamis. The densely populated island of Java holds a possible threat to the social and economic development of the region. The latest strong activity phase of the high risk strato volcano Merapi in central Java commenced in October through November 2010 caused more than 140 casualties. The previous activity phase in May and June 2006 resulted in a estimated lava dome growth of 2.3 million cubic meters (Center of Volcanology and Geological Hazard Mitigation, Yogjakarta), increased seismicity and several pyroclastic flows on its southern flank. During this activity phase of the Merapi, a Mw=6.3 earthquake on 2006 May 26 caused more than 5000 casualties and had its source in the coastal area at shallow depth.
The driving force behind these events is the subduction of the oceanic Indo-Australian lithospheric plate beneath the Eurasian plate. The Sunda Arc curves along the islands of Sumatra and Java with a total length of more than 5000 km. In order to understand the different subduction zone processes, their style and variation along the Sunda Arc, previous investigations from southern Sumatra and western Java (GINCO project) and the most recent investigations (SINDBAD project) off eastern Java and Lombok are presented (Schlueter et al. (2002); Kopp et al. (2006); Planert et al. (2010); Shulgin et al. (2009)) in this chapter. Finally the MERAMEX experiment, conducted off central Java, introduces the aims of this study.

In the framework of subduction zones, alternating phases of tectonic erosion and accretion exist. The style of subducting systems is not static and changes with time, depending on geological and tectonic factors that control the subduction such as convergence rates, seafloor roughness and sediment supply in the trench (Clift and Vannucchi, 2004).
Following previous studies, Clift and Vannucchi (2004) have classified subduction zones into three distinct types: accretionary, erosive and intermediate. Accretionary margins (Fig. 1.1 A), e.g. Cascadia (Hyndman et al., 1990) or Nankai Trough (Taira et al., 1992), are characterized by accreted and underplated trench sediments, often related to mud diapirism and gas hydrate zones. The

CHAPTER 1. INTRODUCTION

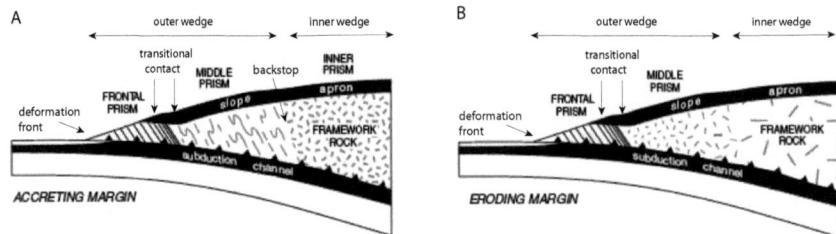

Figure 1.1: Two distinct types of subduction zones (modified after von Huene et al. (2009)) are presented in this sketch. A accretionary , B erosive margins. Please refer to a detailed description in the text.

western Sunda Arc is considered to show a "classical" accretionary structural style. Investigations on this margin advanced the models on sediment accretion and on the evolution of forearc structures (Karig et al. (1979); Curray (1989); Huchon and Le Pichon (1984)). The high ratio of accreted to subducted sediment along Sumatra determines one of the world's largest accretionary wedges. Frontal accretion is the dominant mode off mass transfer along the western Sunda Arc (e.g. Moore et al. (1980); Schlueter et al. (2002)) where only a low percentage of sediment is underplated or subducted, similar to the current situation along the Cascadia margin (Davis and Hyndman, 1989). Offshore western Java, only approximately 1/3 of the current trench fill is subducted and passes the frontal accretionary prism in a subduction window (Kopp et al., 2002).

Erosive subduction zones (Fig. 1.1B), such as Peru and northern Chile (eg. von Huene and Lallemand (1990); Sallares and Ranero (2005)), are tagged by steep trench slopes, a small frontal prism and erosion of upper plate material at the base of the upper plate. Clift and Vannucchi (2004) state that effective tectonic erosion is controlled by collision of bathymetric elevations, and occurs at the front of the margin or along the base of the upper plate. Erosive margins do not form forearc wedges with slope gradients of the upper plate less than 3°. Additionally, erosive margins only exist in regions where the orthogonal convergence rate is greater than 6.3 cm/y (Clift and Vannucchi, 2004).

Intermediate subduction implies that accretionary processes do not exclude erosional processes as suggested by Le Pichon and Henry (1992). Examples from intermediate accretionary wedges (Northern Japan) illustrate frontal accretion and basal erosion in the same wedge.

In order to correctly characterize the structural units of the upper plate, a brief introduction follows based on Wang and Hu (2006) and von Huene et al. (2009) (Fig. 1.1 B). Originating from the deformation front into the landward direction the first unit is the frontal prism, which is composed of actively deforming sediments accreted from the lower subducting plate and slope sediments from the upper plate (Fig. 1.1). A frontal prism is a contractional structure like a small accretionary wedge with landward-dipping reflections that indicate tectonic thickening by imbrication (von Huene et al.,

2009). The 5 - 30 km wide frontal prism can be found at erosive and accretionary margins and merges up slope into an older and more consolidated middle prism by a 1 - 5 km wide transitional contact zone or is separated by a fault zone. The 40 - 100 km wide middle prism consists of decomposed and fractured framework rock with reduced seismic velocities at erosional margins. At accretionary margins consolidated and rigid sedimentary rocks, developed from the frontal prism by imbrication form the middle prism (von Huene et al., 2009). The middle prism is bounded landward by a backstop to the inner prism, which consists of igneous or metamorphic basement and lithified sedimentary rocks (von Huene et al., 2009). The upper plate is covered by an apron of sediments with a thickness from a few meters to 5 km (Clift and Vannucchi, 2004). In this work I use the terminology introduced by Wang and Hu (2006) where the outer wedge represents the frontal part of the forearc overlying the velocity-strengthening portion of the megathrust. The inner wedge is less deformed and covers the velocity-weakening part or seismogenic zone (Fig. 1.1).

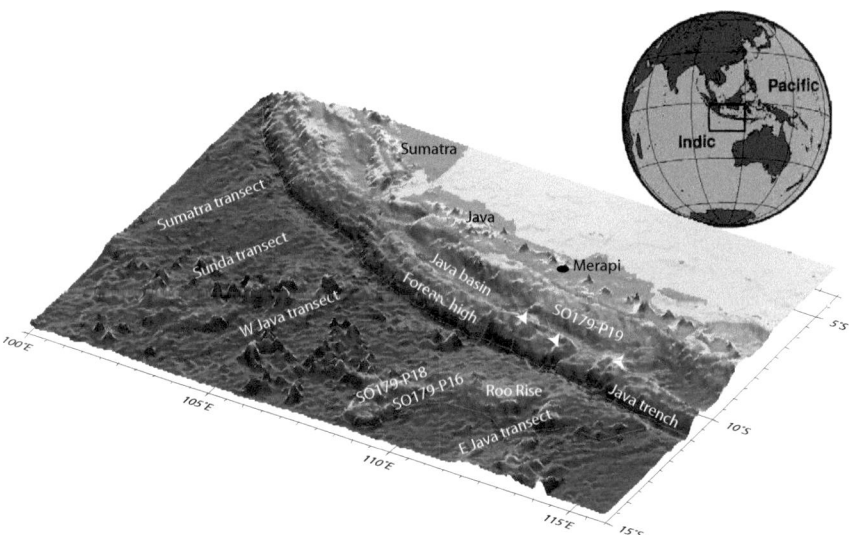

Figure 1.2: MERAMEX Investigation area. The GINCO transects from 1999 (RV SONNE cruises SO137 and SO138) are located at southern Sumatra to western Java (thin red). The SO179 MER-AMEX transects recorded in 2004 are south of central Java. The Merapi volcano (black circle) is positioned in the elongation of the MERAMEX transect. The dashed white line outlines the dimensions of the Roo Rise. The central Java trench retreats in landward direction. Isolated topographic highs along the forearc high (white arrows) indicate a change in the tectonic regime. The most western SINDBAD transect from 2006 (RV SONNE cruise SO190) is situated in eastern Java (thin red).

The sedimentary cover on the Indo-Australian plate along the Sunda Arc decreases with increasing distance from its source, the Ganges-Brahmaputra-System, from west to east. About 1.3 km of trench fill is found off western Java (Kopp et al., 2002). Offshore central and eastern Java, the trench is virtually empty and oceanic basement is exposed. The decrease in sediment supply correlates with the size of the accretionary wedge and the forearc high, which is larger in the north-western portion along the Sumatra segment (Fig. 1.2), where sediment influx on the incoming plate is greater than off Java and trench sediment fill reaches several kilometers (Klingelhoefer et al., 2010). Along the section of southern Sumatra to western Java tectonic accretion with a well-developed accretionary wedge is observed (Kopp et al., 2002). At 109°E a change in the subduction style suggests the domination of erosive processes offshore central Java (Kopp et al., 2006).

The incoming oceanic plate off Java is dotted by numerous topographic highs (Masson et al., 1990), identified as a broad band of seamounts (Christmas Islands Seamount Province) oriented in a W-E direction (Fig. 1.2). Side-scan sonar investigations off eastern Java mapped 10 seamounts with a diameter of 10 - 60 km and a height of more than 1500 m. These seamounts are in different stages of subduction (Masson et al. (1990); Shulgin et al. (2010)). When topographic highs interact with the frontal prism, mass wasting and slumping occurs.

The main topographic unit entering the Java trench comprises the Roo Rise south of central Java (Fig. 1.2), which is interacting with the trench as it is largely responsible for the observed trench retreat by approximately 50 to 60 km from a normal smooth continuation of the trench line (Fig 1.2) between longitudes 109°E and 115°E. It is a little investigated oceanic plateau, which is approximately 2 - 2.5 km higher than the surrounding seafloor and irregular in shape. Here, older refraction seismic profiles indicate a thickened crust with an average value of 11.5 km and a maximum of 16.4 km (Ghose et al., 1990). The absence of a free-air gravity expression of the relief of the Roo Rise suggest that this plateau is a compensated feature supported by a low-density root. The low-density root should be more buoyant than the surrounding material and therefore difficult to subduct. Subduction of lighter oceanic rise material has been correlated to uncommon features in subduction zones, e.g. outer arc uplift, modification of slab dip, discontinuity of the volcanic arc and lack of moderate to large earthquakes (Ghose et al. (1990); Newcomb and McCann (1987)).

1.1 Sunda Arc: Geodynamic setting

The Sunda subduction system evolved after the Eocene collision of India with Eurasia (Fig. 1.3) and is active since middle Tertiary (Hamilton, 1988). Half of the northward motion of India is being accommodated in continental underthrusting at the Tibetan plateau and by compressive thickening of the entire continental crust (Hamilton, 1979). The other half is compensated by the eastward motion of China, obliquely away from the advancing southern continent.

1.1. SUNDA ARC: GEODYNAMIC SETTING

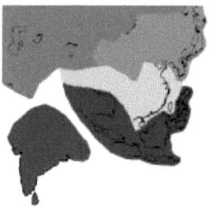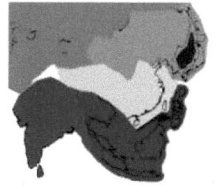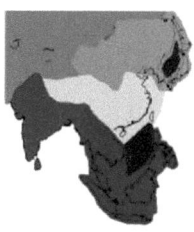

Figure 1.3: Collision of India with Eurasia. India (red) has fractured the plate of east Asia into several subplates (violet, yellow, blue), which have been pushed far to the east and southeast as the collision has progressed. The South China Sea opened as Borneo moved away from China. The sequence starts at 60 million years ago. India pushes blocks of the Eurasiatic plate eastward. Continental margin rocks smear along the margin of Southeast Asia, while the northern part of India thrusts beneath Tibet. (Fig. modified from University of Wisconsin - Green Bay).

The resulting escape tectonics in southeast Asia includes a 10° clockwise rotation of tectonic units in Indochina and Indonesia, as implied by the plasticine indentation experiments of Tapponnier et al. (1982). The trench parallel shear is absorbed by transpressive deformation of the Eurasian plate. The partitioning of oblique plate convergence results in thrust and strikeslip motions.

The oblique collision results in the evolution of strike-slip fault systems like the Sumatra fault (Sieh and Natawidijaja, 2000) and possibly the Mentawai fault zone offshore Sumatra (Diament et al., 1992) (Fig. 1.4). Due to the rotation of Sumatra with respect to Java pull-apart basins developed along western Sunda Strait (Semangka Graben). However, the strongest curvature of the Sunda Arc occurs at the Sunda Strait and forms the transition from frontal to oblique subduction (Malod and Kemal, 1996) or even represent the boundary of two different geodynamic settings (Ghose et al. (1990), Kopp et al. (2001)).

The convergence rate of 6.7cm/yr in a 11°N direction is almost orthogonal to the trench offshore Java and is well determined by GPS measurements (Tregoning et al., 1994). The age of the oceanic crust increases from west to east with 96 Ma off western Java to 135 Ma (Jurassic) off eastern Java (Moore et al., 1980). The dip angle of the downgoing plate increases from 5° off southern Sumatra to approximately 7° off western Java (Kopp et al., 2002).

The deep sea trench along the northern and central Sunda Arc off Sumatra and western Java displays a flat morphology, whereas off eastern Java a V-shaped structure is found (Ganie et al., 1987), implying a reduced sediment supply into the trench. Side-scan sonar data acquired during the 1980's show only isolated small sediment ponds in a generally sediment starved trench. These local accumulations of sediment result from erosive processes related to the collision of seamounts with the deformation front and do not have their origin on the oceanic plate (Masson et al., 1990). The deformation front along this segment of the margin is irregular in shape, related to the Roo Rise subduction.

CHAPTER 1. INTRODUCTION

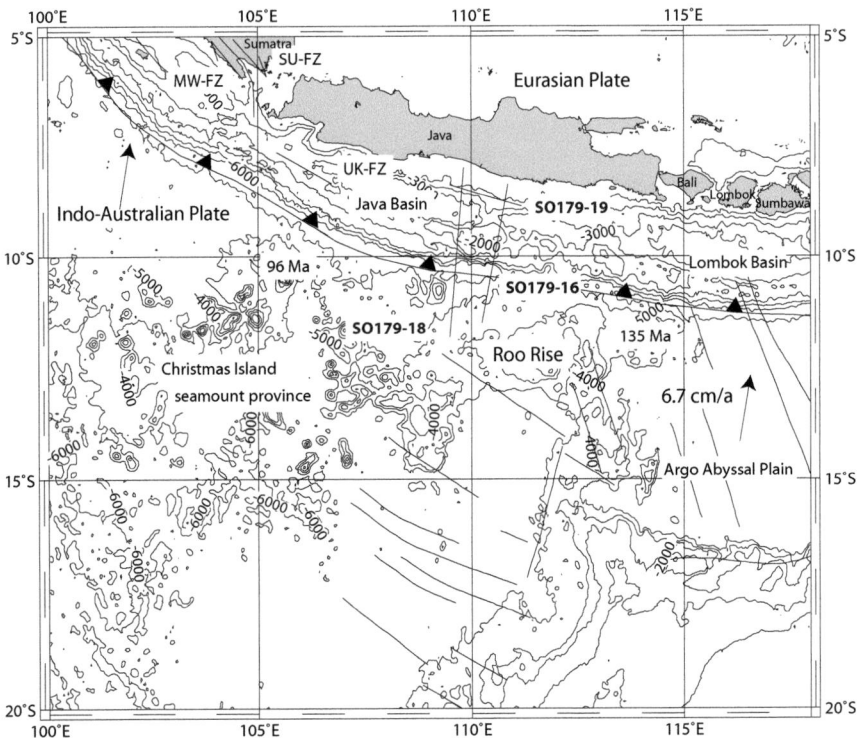

Figure 1.4: Tectonic setting. At the Sunda Arc the Indo-Australian oceanic plate subducts beneath the Eurasian plate with a convergence rate of 6.7 cm/a in a 11°N direction. The crustal age increases from 96 Ma off western Java to 135 Ma off eastern Java. The oblique subduction off Sumatra results in the evolution of fault systems: MW-FZ=Mentawai, SU-FZ=Sumatra, and UK-FZ=Ujong-Kulon fault zone.

1.2. SEISMICITY

Bending related trench parallel faults are found on the oceanic plate. The distance between the fault zones is about 2 - 10 km, the fault lengths of up to 60 km are also commonly recognized (Masson et al., 1990). The vertical height of these faults range from 100 m to 500 m in the vicinity of the trench (Ganie et al., 1987). The plate bending related faulting has been proposed to be the cause of hydration processes in the oceanic lithosphere of the outer rise region (Peacock (1990); Kirby et al. (1996); Jiao et al. (2000)). Recent studies imply that crustal and upper mantle hydration of the oceanic lithosphere occurs in the near-trench setting of the outer rise (Ranero et al. (2003)) and results in serpentinisation of the upper lithosphere. The oceanic crust offshore Lombok is almost devoid of sediments and has an average thickness of 8.6 km/s. The presence of mantle-penetrating faults and a significant mantle serpentinisation determines reduced crustal and mantle velocities of 7.4 - 7.9 km/s in the uppermost 2 km beneath the Moho. This corresponds to an onset of normal faulting within 40 km seaward of the trench (Planert et al., 2010). Carlson and Miller (2003) predict 15 % serpentinisation for most subduction zones from which hydration processes have been postulated.

1.2 Seismicity

Subduction related earthquakes are the source for tsunamis and approximately 90 % of the seismic moment is released along subduction zones. The seismogenic behavior is dependent on the width to the seismogenic coupling zone between the upper plate and the lower plate. The seismic moment M_0 is the product of the total fault area A, the rigidity μ (depends on the material of the fault zone) and the displacement D across the fault: $M_0 = \mu A D$. A larger fault area therefore determines a larger earthquake. The coupling zone is limited by the updip and downdip limit of rupture and is controlled by thermal properties and structural features. The change in frictional properties in the thrust at temperatures of 100 - 150 °C is associated with the updip limit (Oleskevich et al., 1999). The intersection of the thrust with the forearc mantle coincides in many subduction zones with the downdip limit of thrust earthquakes and can be explained by aseismic hydrous minerals present in the mantle. Water trapped in the oceanic crust is released due to higher pressures and temperatures at increasing depths. The dehydrated water migrates into the overlying forearc mantle, forming serpentinite and brucite. Dry and strong rocks become weak and hydrous. These weak rocks do not support seismogenic slip and limit the size of the rupture zone and therefore the seismic moment M_0 of an earthquake (Hyndman et al., 1997).

The Sunda Arc subduction zone has the potential for destructive megathrust earthquakes. The most devastating catastrophic event was the 2004 December 26 earthquake with a magnitude > 9 off Northern Sumatra. In contrast to the Sumatra segment, smaller earthquakes with magnitudes < 8 occurr offshore Java. These earthquakes occur near the trench axis and are often related to seamount subduction. The devastating tsunamigenic potential of the Java subduction zone was released in

CHAPTER 1. INTRODUCTION

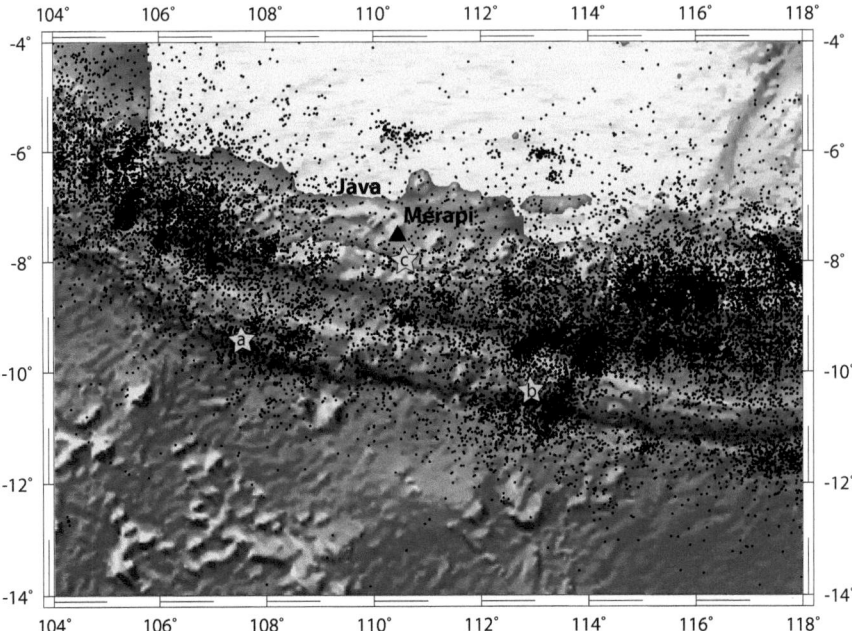

Figure 1.5: Epicenter distribution of earthqakes from 1990 to 2007 (NEIC catalogue) with magnitudes ≥ 1 and ≤ 9. The yellow stars indicate earthquakes described in the text. a: 2006 July 17, Mw=7.7; b: 1994 June 02, Mw=7.8; c: 2006 May 26, Mw=6.4

a thrust earthquake on 1994 June 02 (Fig. 1.5) with a magnitude of Mw=7.8 (Dziewonski et al., 1995). The tsunami killed more than 250 people and had run-up heights reaching 14 m (Tsuji et al. (1995)). Abercrombie et al. (2001) stated that the Java earthquake was caused by a slip over a locked seamount.

The dip angle of the subducting slab increases from west to east, which coincides with a varying maximum hypocentral depth of 200 km beneath Sumatra to 670 km beneath Java (Schoeffel and Das, 1999). This is due to an increasing age of the subducting plate and higher convergent rates compared to the Sumatra segment (Kirby et al., 1996).

The seismicity distribution along the Sunda Arc is not homogeneous (Fig. 1.5). Between 108°E and 111°E along the central Java segment a so called "seismic gap" with a significantly reduced seismic activity can be recognized. Grevemeyer and Tiwari (2006) used thermal models and structural constraints derived from seismic and gravity data to explain the seismogenic behavior in the Sunda subduction zone. They suggest that a shallow serpentinised mantle wedge limits the seismogenic coupling zone off Java, which determines a lower potential for megathrust earthquakes like the 2004 December 26 earthquake off Sumatra. On 2006 July 17 the most recent devastating earthquake occurred with a magnitude of Mw=7.7 and 900 casualties due to a 2 m high tsunami wave (Fig. 1.5). Like previous tsunami earthquakes, the Java event had an unusually low rupture speed of 1.0 - 1.5 km/s, and occurred near the up-dip edge of the subduction zone thrust fault (Ammon et al., 2006). The rupture propagated up-dip and slowly, with a relatively long duration of $\approx$185 s. Most large aftershocks involved normal faulting. The rupture propagated 200 km towards the trench (Ammon et al., 2006). A slow rupture propagation and a long rupture duration are typical for tsunami earthquakes (Polet and Kanamori, 2000).

The magnitude Mw=6.4 2006 May 26 earthquake was located in the densly populated district of Yogjakarta, in the coastal area at shallow depth of the MERAMEX investigation area. It was not a subduction triggered earthquake, but rather displayed strike-slip characteristics associated with a fault system and could be related to the increased activity of the Merapi volcano. The aftershock earthquake distribution correlates with a northwest-southeast oriented low-velocity zone, revealed with tomographic studies by Wagner et al. (2007).

1.3 GINCO (1999):
Investigations along southern Sumatra to western Java

Off western Java reflection profiles were collected during the RAMA cruise of RV Thomas Washington in 1980 (Benaron, 1982), which were complemented by the GINCO dataset in 1999 (Kopp et al., 2002). In late 1998 and early 1999 multichannel seismic- and wide-angle data were collected with RV SONNE (SO137 & SO138) in the south Sumatra to west Java region (Fig. 1.2). The GINCO (Geoscientific Investigations on the active convergence zone between the east Eurasian and Australian plates along Indonesia) project was initiated to investigate the large scale structure of the continental margins of Sumatra and western Java (Schlueter et al., 2002). The focus of this project lies on the identification of the different geologic-tectonic features associated with the subduction-accretion processes. Kopp and Kukowski (2003) could map exactly the backstop system geometry and margin segmentation. The results of the Sumatra, Sunda and western Java transect of the GINCO project (Fig. 1.2) will be gathered for a comparison with the MERAMEX study interpretation.

The incoming igneous oceanic crust shows a normal velocity-depth function according to Pacific oceanic crust samples older than 29 Ma (White et al., 1992). Due to the lack of Pn phases, the oceanic mantle velocity was set to 8.0 - 8.1 km/s. The sediments reach a thickness of 1.5 km in the trench and are pushed toward the toe of the accretionary wedge. The accretionary domain shows a chaotic seismic character, indicating deformation. Sediments are accreted and underthrust beneath the accretionary domain which results in uplift and intense faulting (Kopp et al., 2002).

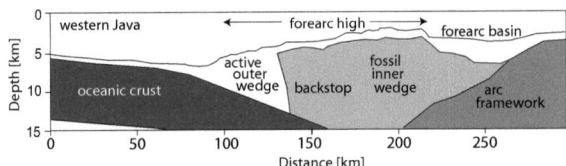

Figure 1.6: Tectonic segmentation of the western Java forearc high. This figure is based on Kopp et al. (2002)

Along the margin from southern Sumatra to western Java the forearc high can be distinguished between two types of accretionary wedges (Fig. 1.6). The fossil inner wedge is of Paleogene age and forms the forearc high. It builds a backstop structure for the Neogene to recent aged active outer wedge (Kopp and Kukowski, 2003). The outer wedge is characterized by landward-dipping imbricate thrust sheets with a width of approximately 4 - 6 km (Kopp et al., 2009). Imbricate thrusting determines an arcward thickening of the accretionary wedge. This is a common process for accretionary subduction zones where the wedge becomes progressively more consolidated and cemented towards the arc (Kopp et al., 2009). Therefore the lateral growth of the inner wedge is mainly effected by addition of material from the outer wedge (von Huene et al., 2009).

The forearc high evolved as a consolidated and lithified dynamic backstop. The definition "backstop" is independent of the upper plate composition (Davis, 1996), which changes from continental off Sumatra to oceanic-type off western Java (Hamilton (1979); Grevemeyer and Tiwari (2006)). The material of the forearc high was initially pushed against the arc rock framework during the early stages of subduction resulting in a morphological elevation. It is also called dynamic backstop due to its deformation to adjust its taper. The increase in internal strength from the frontal accretionary prism to the forearc high of the Sunda Arc leads to a shallowing of the surface slope (Kopp and Kukowski, 2003). The stronger material is able to support a narrower wedge while still undergoing stable sliding at its base. A shallowing of the surface slope is common with accretionary wedges, including the Nankai and Barbados accretionary wedges (Lohrmann et al. (2003); Bangs et al. (1990)), and the Alaskan margin (Ryan and Scholl (1989); Fruehn et al. (1999)).

The forearc high basement is assumed to be composed of metamorphosed sediments near the base. The seaward part of the 75 km wide forearc high shows little evidence for deformation, the landward part toward the forearc basin is experiencing ongoing activity (Kopp et al., 2002). At the top of the basement several basins are formed, resulting from the steepening of the old accreted strata now forming the backstop. Sediment ponded in these basins is highly deformed, probably caused by strike-slip motion transferred from the large strike-slip systems associated with partitioning of plate convergence (Hindle, D., pers. communication).

The forearc basement shows a layer of elevated velocities beneath the forearc basin, which is a

oceanic-type P-wave velocity structure. Approximately 5 km of sediments are trapped in the forearc basin. Beneath the shelf higher velocities (> 5.9 km/s) are attributed to rotated basement blocks of metamorphosed pyroclastic sediment, which were identified in the MCS data (Kopp et al., 2002).
The shallow upper plate Moho (approximately 10 km below seafloor; in the following abbreviated as bsf) beneath the forearc basin is supported by the 2D gravity modeling which expects high density material (density of 3.37 g/cm^3) here. A possible source for oceanic-type crust might be remnant fragments of former oceanic crust that have been altered and might be related to the subduction of the former Tethys ocean.

1.4 SINDBAD (2006):
Investigations along eastern Java to Bali and Lombok

SINDBAD (Seismic and Geoacoustic Investigations Along the Sunda-Banda Arc Transition) is a joint German-Indonesian project. The Sunda-Banda subduction zone, which marks the southern limit of the Indonesian archipelago curving along the islands of Java, Bali, Lombok and Sumbawa (Fig. 1.4), is the eastward expression of the western and northwestern Sunda margin. The style of subduction varies from an oceanic-island arc type along the eastern Sunda margin to continental-island arc collision along the Banda margin. The morphologic variations of the study area include the Roo Rise, the Argo Abyssal Plain, and the continental lithosphere of Australia (Fig. 1.4). The investigations focus on the evolution of the overriding plate by imaging the deep and shallow crustal structures using multichannel seismic, magnetic, and gravity data, as well as analyses of seismic wide-angle data collected through OBS, swath bathymetry end echo soundings (Mueller et al., 2008).
The seismic analysis conducted in late 2006 during cruise SO190 of RV SONNE supplies information on the subduction system input to quantify the mass transfer from the deep sea trench to the deeper forearc. The cruise SO190 was comprised of two legs: Leg 1 was conducted by the Federal Institute of Geosciences and Natural Resources (BGR), and was dedicated mainly to seismic multichannel data acquisition, complemented by gravity and magnetic studies. Leg 2 was conducted by IFM-GEOMAR, with the main focus on wide-angle seismic profiling, complemented by additional seafloor mapping potential field measurements.
Shulgin et al. (2010) investigate the most western wide-angle seismic profile of the SINDBAD project, which is located off eastern Java (Fig. 1.2), by applying a P-wave velocity model, additionally complemented with gravity modeling. The crustal thickness ranging from 12 - 18 km was resolved by a joint refraction and reflection 2D tomography (Korenaga et al., 2000) in the central portion of the Roo Rise. The structure of the upper crust of the downgoing oceanic plate shows a high degree of fracturing in its top section, which is clearly visible in the multichannel seismic data down to 2 km (Shulgin et al., 2010). It is suggested that the crust is cut by faults even to greater depths, which

is indicated by low mantle velocities of 7.5 - 7.9 km/s. The trench is devoid of sediments, except for local sediment ponds associated with original seafloor fabric (Planert et al., 2010). The deformation front is retreated in landward direction due to the strong influence of the subducting Roo Rise and the rough topographic relief of the oceanic crust. The depth of the mainly undisturbed forearc basin is approximately 1500 m, which is less than the well-developed forearc basin to the east where the sediment fill reaches 3 km (Planert et al., 2010). The forearc high is dotted by a number of uplifted isolated highs, like off central Java. Shulgin et al. (2010) provide two possible interpretations for the uplifted forearc high of the eastern Java profile. The first variant suggests a buoyant fragment of the oceanic plate located southward of the static backstop, which was detached from the downgoing plate and stacked over the static backstop. The observed forearc high corresponds to the area of maximum stacking of a rough oceanic plate, dotted by numerous elevated features. The second variant suggests a bending related uplift of the forearc edge, caused by the presence of stacked fragments of the oceanic origin below it. This was also observed in the Lesser Antilles subduction zone (Bangs et al., 2003).

1.5 MERAMEX project (2004) and motivation (this study)

The central segment of the Sunda Arc from southern Sumatra to western Java is well-studied (GINCO) and the most eastern Sunda Arc segment was recently investigated by the SINDBAD project (chapters 1.3 and 1.4). The MERAMEX project is located in between these study areas off central Java.
Bollinger and de Ruiter (1976) evaluated seismic profiles across the forearc basin south of central Java and found an unconformity, where younger sediment onlaps the landward margin of the forearc basin. Curray et al. (1977) described thickened oceanic crust from wide-angle seismic data with sonobuoys. Moore et al. (1980) presented single-channel reflection profiles south of central Java and investigated the trench slope and slope basins. They emphasized the steep trench slope and the deformation of the forearc basin strata. A detailed subduction structure could not be resolved due to insufficient seismic energy penetration using sonobuoys and explosives as seismic source. Furthermore, only first arrivals were used which were often masked by later arrivals.
To close the gap of poorly resolved data off central Java the marine components of the MERAMEX (Merapi Amphibious Experiment) project were established in 2004 to investigate the detailed crustal structure of the subduction zone off central Java in association with the activity of Merapi volcano. Three wide-angle seismic profiles were acquired off central Java in autumn 2004 aboard RV SONNE (SO179 cruise), which are presented in this study. The combined onshore and offshore passive experiment commenced in spring 2004. Onshore, more than 100 seismological stations were deployed in a dense network around Merapi volcano (Fig. 1.7),

which recorded data for a period of more than 150 days. Based on this dataset, the crustal structure beneath Merapi volcano was investigated with active and passive tomographic studies by

1.5. MERAMEX PROJECT (2004) AND MOTIVATION (THIS STUDY)

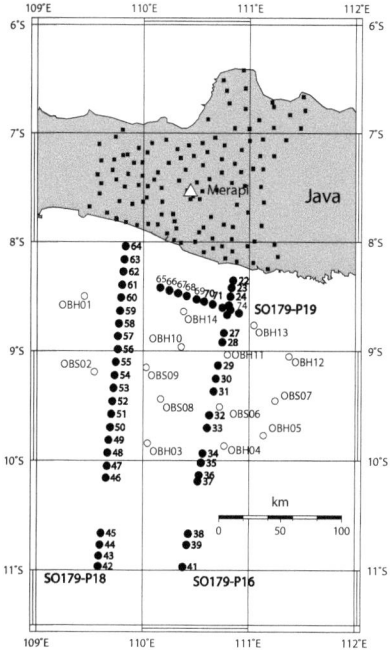

Figure 1.7: Station distribution of the MERAMEX experiment. Onshore more than 100 temporary seismological stations were installed around Merapi volcano (black squares). Offshore a temporary seismological OBS/H network was deployed (open circles). The three wide-angle and reflection seismic profiles SO179-P16 to SO179-P19 were covered with 53 OBS/H stations.

Koulakov et al. (2007) and Wagner et al. (2007).
Offshore, on the SO176 cruise of RV SONNE, a network of fourteen ocean bottom seismometers and -hydrophones (OBS/OBH) stations was deployed to collect seismological data (Fig. 1.7). These earthquakes in combination with the data from the temporal onshore seismological network were used to investigate the local seismicity in the light of tsunamigenic hazards and the seismic gap (Chapter 1.2) off central Java. On SO179 cruise the seismological stations were recovered offshore and the three wide-angle seismic lines were shot; these data were complemented by hydrosweep, gravity and magnetic measurements.
This study is focused on the evaluation of the offshore wide-angle seismic- and gravity data. The methods applied in this study include forward modeling of first and later arrivals of P-wave traveltimes with raytracing (Luetgert, 1992). The aim is to model the detailed P-wave velocity structure and the

possible development of the subduction system from accretionary to erosive tectonics. In alternation on the raytracing forward models a tomographic inversion is applied and supports the Moho depth from the forward modeling by a joint inversion of reflected and refracted first arrival traveltimes with the tomographic algorithm compiled by Korenaga et al. (2000). This forward and inverse approach was applied alternately until a minimum traveltime misfit and a minimum forward and inverse model difference was reached. A forward model of the gravity data lead to additional constraints on the P-wave velocity data.

The determined models off central Java provide information on the incipient subduction of the oceanic Roo Rise plateau and its influence of forearc kinematics. The segmentation of the forearc high along the Sunda Arc and the main differences to the GINCO and SINDBAD transects are compared and linked in this study. The resulting images document the changes in the tectonic and morphological regime offshore Java.

Chapter 2

Seismic data

Three wide-angle and reflection seismic profiles were acquired on RV SONNE cruise SO179 south off central Java. Two sub-parallel profiles were shot perpendicular to the trench and are about 100 km apart. Profile SO179-18 at 110°E starts on the oceanic crust, trends to the north and ends after a length of 340 km on the shelf area. The eastern profile SO179-16 is located at 111°E and is 360 km long (Fig. 1.7). Station 40 of profile SO179-P16, located on the outer high, could not be recovered. Station OBH63 had a power failure and the data could not be recovered properly from flash disk and could not be used for data modeling. Stations 24 - 22 located at the shelf break of this profile showed a low signal-to-noise ratio and could not be used for modeling. The reflection streamer data of profile SO179-P18 were corrupted due to a leakage. Therefore the reflection seismic section could not be used to interpret this line. The third 200 km long profile SO179-19, consisting of 11 stations, was positioned parallel to the trench in the shelf area and crosses the two dipping profiles (Fig. 1.7). Seismic reflection data were recorded along the entire profile, whereas wide-angle seismic data were acquired from profile-km 80 - 200 km. Spacing between the stations was approximately 12 km.

2.1 Data acquisition

The wide-angle seismic signals were recorded with a sampling interval of 5 ms. Due to drifting of the OBH/S stations during sinking, the station positions may have been dislocated. Two short cross profiles were shot over the stations, and were only used to verify their exact position on the seafloor. With a maximum operation depth of 6000 m, no OBS station could be deployed in the deepest part of the trench, which determined a short profile gap. Most of the wide-angle data are of high quality with some stations showing phase information over distances up to the whole profile length (Fig. 2.1).

For the seismic profiles, three airguns (32 l Boltguns) with a dominant frequency of 6 - 8 Hz were used. They were towed 60 m behind the vessel at a depth of 7 - 8 m, and operated at 145 bar. The shot interval was 60 s with a vessel speed of 4 knots resulting in a shot spacing of approximately 120 m. In addition, a ministreamer was used during the wide-angle data acquisition to record vertical incidence reflection data from the upper sedimentary sections and the basement reflector. The streamer consists of four 12.5 m active sections with 25 hydrophones spaced at 0.5 m. The streamer was placed about 8 m behind the vessel. Individual hydrophones are omnidirectional and have a flat frequency response from 10 - 1000 Hz.

2.2 Data processing

To improve the temporal resolution of the seismic data a deconvolution was applied. The amplitude spectra of the seismic traces vary with time and offset, hence the deconvolution must be able to follow these variations. Input for the deconvolution process are the raw data. Several recordings were influenced by a DC shift, therefore a 1 - 3 Hz high-pass minimum delay Kaiser frequency filter with 60 dB attenuation between the pass and reject zone was applied. With a predictive length of 140 ms and an operator length of 300 ms a multitrace Wiener deconvolution was chosen which is a compromise between temporal resolution and signal-to-noise ratio.

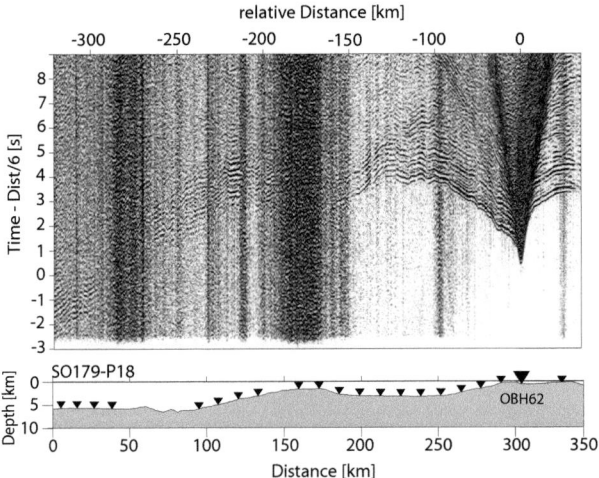

Figure 2.1: Wide-angle record section of OBH62 (profile SO179-18) located on the Javanese shelf in very shallow water depth (632 m). The first arrivals can be traced over the complete profile length. All following seismic record sections are displayed with a reduction velocity of 6 km/s.

2.2. DATA PROCESSING

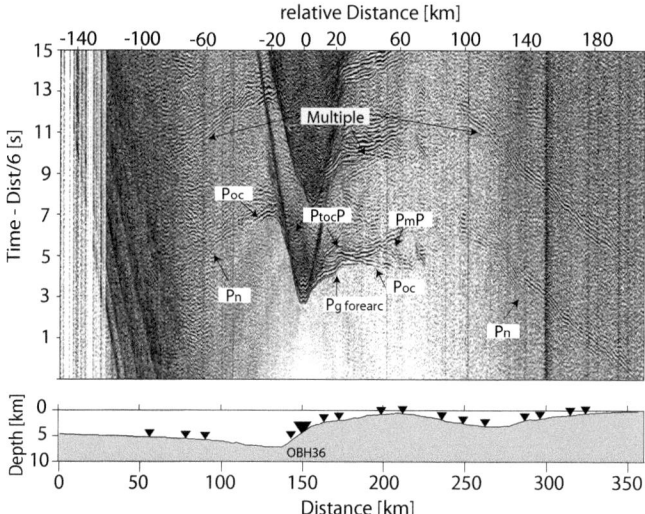

Figure 2.2: Wide-angle record section of OBH36 (profile SO179-P16). This data example shows typical phases from stations near trench locations on the upper plate. The phases are denoted as following: Pn= mantle phase (oceanic and margin wedge mantle respectively); PmP= Moho reflection, Poc= refraction from oceanic crust, PtocP= reflection from the top of the oceanic plate, Pg forearc= refraction from the forearc high.)

After deconvolution an offset- and time variant Ormsby filter with minimum delay characteristics was applied. As the seafloor depth changes along the seismic lines, each trace was statically corrected to a fixed seafloor traveltime of 11 s based on the waterdepth before filtering. Subsequent to wide-angle data processing, the traveltime curves of all stations were picked. In total approximately 30.000 P-wave traveltime picks (including refracted and reflected phases) of 44 stations were used for the modeling process.

The first arrivals were picked in the near offset range ($\leq$ 30 km) with uncertainties of less than $\pm$50 ms. At greater offsets ($\geq$ 30 km) the uncertainties reach approximately $\pm$100 ms. The large pick uncertainties are due to a lower signal-to-noise ratio with increasing offset. The phase Poc (oc denotes oceanic crust) (Fig. 2.2) is refracted through the oceanic crust and is used to identify the P-wave velocities in the oceanic crust. All stations located in the near trench area at the upper plate trace the top of the subducted plate as a strong reflection PtocP (toc denotes top of the oceanic crust) and refraction Poc (Fig. 2.2). Moho PmP reflections are identified by high amplitudes on stations on the outer rise from both dip-profiles and from near trench stations on the upper plate. Phases from the plate boundary are present with strong reflections to station OBH36 at profile-km 180 (Fig.

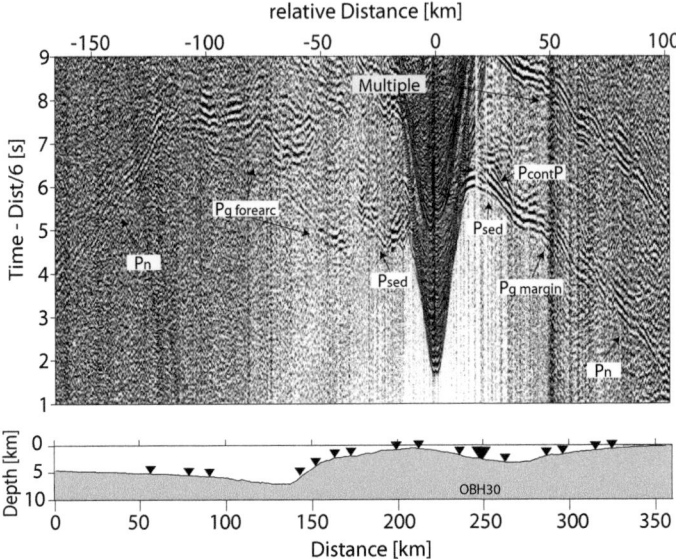

Figure 2.3: Wide-angle record section of OBH30 (profile SO179-P16). This data example shows typical phases from stations on the forearc basin locations of the upper plate. The phases are denoted as following: Pn= mantle phase (oceanic and margin wedge mantle respectively; PcontP= reflection from continental Moho, Psed= shallow sedimentary refractions, Pg margin= refractions from the margin wedge, Pg forearc= refraction from the forearc high.)

2.2). PmP reflections could only be identified on profile SO179-P18 from stations located on the outer rise. Oceanic mantle Pn phases could often be traced over a distance of 100 km by stations positioned on the oceanic crust.

Phases from the margin wedge (upper plate) (Pg margin) are denoted as 'intracrustal' refractions and reflections. The continental Moho depth could be modeled with PcontP phases (Fig. 2.3). The intracrustal phases of SO179-19 are of high quality and could be identified over the total profile length of 200 km.

Chapter 3

Modeling the seismic P-wave velocities from wide-angle data

In this chapter the seismic P-wave velocity models of three wide-angle seismic profiles will be investigated by applying seismic forward modeling, followed by an inverse step. In this study both methods are used alternately to supplement each other.

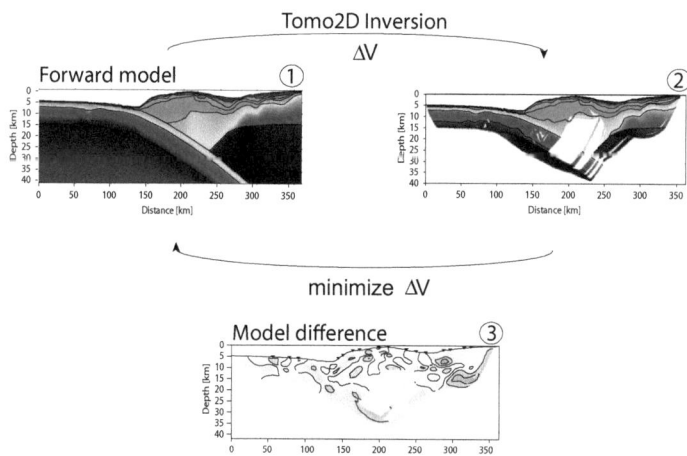

Figure 3.1: Modeling strategy for this study. The forward model (1), based on interactive 2D raytracing provides the input model for the tomography (2,inverse step). The forward model is updated with the model differences δV (3). This alternating forward and inverse calculations are repeated until the model differences δV and the misfit between calculated and observed traveltimes is minimized.

20 CHAPTER 3. MODELING THE SEISMIC P-WAVE VELOCITIES FROM WIDE-ANGLE DATA

The idea of the forward modeling is to solve the equation of motion for seismic waves. Rays are traveling through the geological model and the corresponding synthetic traveltimes are compared with the real seismic data. If the fit is within an acceptable level of accuracy, the model can be taken to be a reasonably accurate model of the subsurface. If the traveltime misfit is too large, the model is altered and new synthetic traveltimes are computed. This process continues iteratively until the misfit between calculated and real traveltimes matches the requirements. The opposite of the forward modeling is the inverse approach in which the velocity-depth model is computed from the acquired traveltime data. The inversion is based on the linearized relationship between the traveltime data and the velocity structure **Gm=d**, where **m** is the unknown slowness vector (model), **d** is the traveltime vector (data), and **G** is a matrix whose rows contain path lengths through each model element (grid) for a given raypath (refer to chapter 3.2 for more details).

The forward models provide the input models (Figure 3.1) to the inversion based on the 2D tomography code by Korenaga et al. (2000). The same first arrival traveltime data are used for the forward and inverse step. The resulting inverted velocity-depth models minimize the traveltime misfit after four iterations between the calculated and observed traveltimes. The determined model differences ΔV between forward and inverted (Figure 3.1) velocity models are updated and refined in the forward model. The updated forward model provides a new input model to the inverse step. The advantage of this procedure is the full control and stepwise adjustment of the model structures, which are build in the forward model.

3.1 Forward modeling with raytracing

The interactive 2D raytracing program MACRAY (Luetgert, 1992) was used to create a first forward velocity-depth model of the three profiles. With this technique not only first arrivals but also later refraction and reflection phases could be used to resolve complicated structures in the profiles. The algorithm calculates the propagation of rays within a layer by the stepwise integration of the system of first order differential equations (Luetgert, 1992):

$$\frac{dx(t)}{dt} = V(x,z)\sin\theta$$
$$\frac{dz(t)}{dt} = V(x,z)\cos\theta$$
$$\frac{d\theta(t)}{dt} = \frac{dV}{dx}\cos\theta - \frac{dV}{dz}\sin\theta$$

of x,z and θ, where θ is the ray angle from the vertical. The theoretical rays and their corresponding traveltimes are calculated from a 2D laterally heterogeneous model and then compared to the acquired wide-angle data. The starting model only includes the known bathymetry. In an iterative approach the velocity-depth model is modified until the calculated traveltimes fit the observed data. In a top to bottom approach each single layer was adapted from the sedimentary sections down to the mantle. The upper sedimentary sections were also identified in the reflection seismic data and integrated into the wide-angle model. This procedure was used to gain independent two dimensional velocity-depth models of the three wide-angle seismic profiles. Finally, the three profiles were merged together to obtain three dimensional geometrical constraints of the area. The quality of the model depends on qualitative estimates of phase identification uncertainty (refer to chapter 2).

3.2 Inverse modeling with 2D tomography

In this chapter the forward modeling is complemented and refined alternately by the inverse step to compute the velocity structure of the central Java subduction zone with the given data set of three wide-angle profiles. This chapter briefly introduces the first arrival joint refraction and reflection tomography code Tomo2D (code developed by Korenaga et al. (2000)), followed by an explanation of the detailed model parameterization.

A 2D velocity model is parameterized as a sheared mesh hanging beneath the seafloor (Korenaga et al., 2000). The smooth velocity field within the parallelogram-shaped grid cells is bilinearly interpolated. Nodal spacing can vary in horizontal and vertical directions. A velocity mesh should be finer than the expected velocity variations to avoid any bias caused by a coarse grid parameterization. The forward model based on MacRay is converted to the Tomo2D input format by a unix shell script. All the preliminary information, like velocities and reflector depths provide the input model for the tomography.

The reflector is represented as an array of linear segments whose nodal spacing is independent of that used in the velocity grid (Korenaga et al., 2000). Horizontal coordinates of reflector nodes are fixed so that each node is only updated in the vertical direction after each iteration. The floating reflector formulation determines reflector segments which are independent of adjacent velocity nodes. Therefore Pn rays are necessary to calculate velocities below the Moho.

Forward traveltime calculation takes a hybrid approach. A graph-theoretical method is used to ensure a global optimization, followed by ray-bending refinement to achieve the desired accuracy. The inversion with fine model parameterization is regularized by smoothness and damping constraints on both velocity and reflector nodes. In the joint reflection/refraction inverse problem the inverse solution is not unique and depends on the starting velocity model. This dependency must be assessed by conducting a number of inversions with different starting models (refer to chapter 3.3.1).

The joint inversion of refraction and reflection traveltimes was applied on the western and eastern profiles with PmP mantle (oceanic Moho), PcontP (Moho mantle wedge) and PtocP (reflections from

top of the subducting plate) phases. The quality of reflection phases was sufficient to ensure a good resolution of the reflectors.
In the following a brief introduction of the Korenaga code and the used model parameters are presented.

3.2.1 Forward problem

The forward calculation of rays and traveltimes in Tomo2D utilizes a hybrid raytracing approach, which applies the shortest path method (also known as graph method) (Moser, 1991) in the first step, and thereafter a ray bending method (Van Avendonk, 1998).

3.2.1.1 Shortest path method (SPM)

The SPM tracks the propagation of the whole wavefront and is based on Fermat's principle. It states that raypaths taken between two points can be traversed in a minimum traveltime. In our case the rays follow their shortest paths between network nodes of a sheared mesh. Huygen's principle states that neighboring nodes act as scattering sources. The velocity model is sampled in each node location with a bilinear interpolation. For each connection a weight is computed as numerically integrated slowness along a straight line. The shortest path follows those connections for which the weight sum is smallest. After Fermat this is an approximation of a seismic ray with a minimum traveltime.
Forcing ray paths following connections introduces errors in ray geometry and calculation of traveltimes. Rays travel zig-zag in homogeneous or smooth velocity zones. This introduces longer ray paths (Moser (1991); Fischer and Lees (1993)) and causes an overprediction of velocities. This overprediction of velocities depends on the number of mesh nodes and the count of connections per node (Moser, 1991). Errors are worst in propagation directions which are poorly covered by available connections (eg. Van Avendonk et al. (2001)), accordingly a forward star (Fig. 3.2) was introduced to obtain a good coverage of search directions for available paths. A higher forward star (which is equivalent to a denser grid node sampling) results in a better solution but needs extra computation time. For crustal velocity models, where a vertical velocity gradient usually dominates the horizontal, a star which preferably searches in the downward direction is favorable with respect to an isotropic star (Van Avendonk et al., 2001). SPM provides an initial guess for the ray bending method: ray paths should be close enough to true ones, therefore ray bending will not fail and results in a global minimum. In this study a forward star of 5 x 10 was used.

Finally, the graph method has to be completed with a connection from the shot location to the seafloor node or receiver, because the water column is outside the sheared mesh. Searching all possible ray paths with two possible endpoints, after Fermat's principle the path with minimum traveltime yields the approximated traveltime.

3.2. INVERSE MODELING WITH 2D TOMOGRAPHY

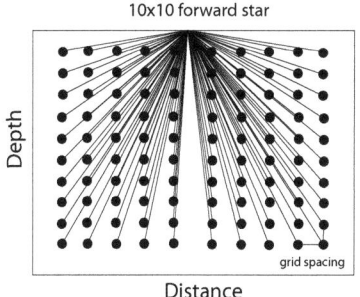

Figure 3.2: Forward star used in the SPM to provide a good ray coverage in all search directions.

3.2.1.2 Ray bending

After SPM a ray path refining with conjugate gradient ray bending technique (Moser et al., 1992) is introduced, where the gradient of traveltime vanishes for a true ray path. The conjugate gradient method iteratively optimizes the calculated traveltimes of the preliminary path, where minimum traveltimes are searched with small geometry perturbation variations along the initial path. The bending algorithm can not diverge because non-zero gradients do not exist. This results in a less traveltime. The iteration is stopped, when the calculated traveltime reaches a threshold.

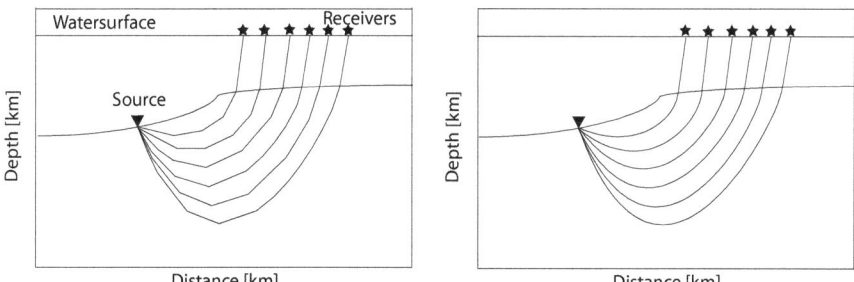

Figure 3.3: Comparison of ray paths in SPM method and after refining the path in ray bending method.

In Tomo2D a conjugate gradient is applied, were rays are parameterized as beta-splines which can express a variety of curves with a small number of control points. After SPM a ray is defined as a number of points connected with straight line segments (polygonal paths) (Fig. 3.3). To avoid wrong ray paths due to low velocity regions a parameterization is used where the spacing of points is adapted to variations in the velocity gradient (eg. Van Avendonk et al. (2001)).

3.2.2 Inverse problem

The Korenaga et al. (2000) method utilizes Fermat's principle to linearize the inversion. Smoothing and damping constraints are used to regularize the system of normalized equations. The linear system is solved using a least-square method LSQR (Paige and Saunders, 1982).
The traveltime along a ray path P is defined as:

$$t_{obs} = \int_P u(r)dr \tag{3.1}$$

with position vector r, the infinitesimal ray length dr and the slowness $u(r)$ at point r. The ray path depends on the solution of the nonlinear problem. The perturbed model $\delta u(r)$ is a function of the initial model $u_0(r)$:

$$\delta u(r) = u(r) - u_0(r) \tag{3.2}$$

Fermat's principle requires a minimum traveltime along a ray path P for an infinitesimal perturbation δu, which results in a small change in traveltime:

$$\delta t_j = t(u + \delta u) - t(u) \approx \int_{P_j} \delta u dr \tag{3.3}$$

The calculation of reflection traveltimes is similarly related to vertical changes in reflector depths and slowness perturbations with:

$$\delta t_j = \int_{P_j} \delta u dr + \frac{\delta T}{\delta z} \delta z(x_j) \tag{3.4}$$

where x_j is the reflection point of the jth ray.
We assume that changes in traveltimes are related to changes in velocities. We can discretize the equations 3.3 and 3.4 for the traveltime residuals with respect to the perturbational model parameters and formulate it as a matrix equation:

$$d = G\delta m \tag{3.5}$$

d describes the n x 1 vector which contains the traveltime residuals, δm is the unknown m x 1 model perturbation vector, and G is the m x n Fréchet derivative matrix (eg. Menke (1989); Toomey et al. (1994)). Partial derivatives with respect to slowness in matrix G are path lengths distributed to the velocity nodes. The depth sensitivity in G is given by reflector inclination, incident ray angle, and slowness in the reflection point (Bishop et al., 1985). The model update vector δm is scaled with model parameters in the original starting model $\delta m' = C^{-1/2}\delta m$. d is scaled with data covariance matrix d'

3.2. INVERSE MODELING WITH 2D TOMOGRAPHY

$= C_d^{-1/2}d$, and the Fréchet derivative matrix is normalized with the relation $G' = C_d^{-1/2}GC_m^{-1/2}$ to avoid possible solution bias towards a model that is characterized by increased levels of heterogeneity at greater crustal depths (Toomey et al., 1994). Ray pick uncertainties are included in the diagonal data covariance matrix C_d. If equation 3.5 is under-determined, smoothness constraints have to be applied to obtain a unique solution. In all smoothing matrices a Gaussian smoothing within one decay length (correlation length) is applied for each perturbational model parameter (Toomey et al., 1994). The correlation lengths vary in horizontal and vertical direction. Due to stronger vertical variations in the vertical Earth's structure, the horizontal correlation lengths are an order of magnitude greater than the vertical ones. Thus vertical C_{Vv} and horizontal C_{Hv} smoothing matrices are applied separately (Korenaga et al., 2000). Each smoothing equation for an individual model perturbation δm_i is normalized by the slowness of the starting model ou_i (Toomey et al., 1994):

$$\delta m_i o u_i^{-1} = \frac{\sum_{j=1}^{m} \beta_j \delta m_j o u_{-1 i}}{\sum_{j=1}^{m} \beta_j} \tag{3.6}$$

The weights β_j decrease with distance from the ith model parameter in a Gaussian distribution:

$$\beta_j = exp\left\{-\frac{(x_j - x_i)^2}{\tau_x^2} - \frac{(z_j - z_i)^2}{\tau_z^2}\right\} \tag{3.7}$$

with τ_x and τ_z as horizontal and vertical correlation lengths for the weights β_j. This ensures that only nodal positions lying within one decay length of the particular model parameter are affected by the spatial smoothing constraints.

The matrix for the inverse problem (corresponding to 3.5) is formulated as following (Korenaga et al., 2000):

$$\begin{pmatrix} d & 0 & 0 & 0 \end{pmatrix} = \begin{pmatrix} L_v & wL_d \\ \lambda_v C_{Hv} & 0 \\ \lambda_v C_{Vv} & 0 \\ 0 & w\lambda_d L_d \end{pmatrix} \begin{pmatrix} \delta m_v \\ 1/w \delta m_d \end{pmatrix} \tag{3.8}$$

where the model vector describes the velocity and depth sensitive components. The weights for the slowness and reflector depth perturbations λ_v and λ_d control the smoothing constraints with respect to the data misfit. The corresponding normalized smoothing matrices for velocity perturbations are C_{Hv} and C_{Vv} and C_d the corresponding smoothing matrix for the reflector depth perturbations. The relative depth sensitivity in the Fréchet matrix is controlled by the depth kernel weighting parameter w. Due to the velocity-depth ambiguity (Bickel, 1990) traveltime data exhibit ambiguities that prevent the resolution of a time anomaly into reflector structure and media velocity. It is possible with the inversion of Korenaga to assess many values for w as a single controlling parameter to explore the possible solution space (Korenaga et al., 2000).

The sparse matrix solver LSQR by Paige and Saunders (1982) calculates equation 3.8 for δm.

26 CHAPTER 3. MODELING THE SEISMIC P-WAVE VELOCITIES FROM WIDE-ANGLE DATA

Damping is introduced to keep the inversion in the region of linearity. If the starting model is far from the final model the calculated ray paths induces large model updates and the inversion may become unstable. Equation 3.8 is updated with some damping constraints:

$$\begin{pmatrix} d & 0 & 0 & 0 & 0 & 0 \end{pmatrix} = \begin{pmatrix} G_v & wG_d \\ \lambda_v C_{Hv} & 0 \\ \lambda_v C_{Vv} & 0 \\ 0 & w\lambda_d L_d \\ \alpha_v D_v & 0 \\ 0 & w\alpha_d D_d \end{pmatrix} \begin{pmatrix} \delta m_v \\ 1/w\delta m_d \end{pmatrix} \tag{3.9}$$

where D_v is the velocity damping and D_d is the depth damping matrix. These matrices can be derived from a penalty function for the magnitude of model perturbation (Van Avendonk, 1998). The weighting parameters α_v and α_d control their particular strength.

It is necessary to stay linear, therefore each inversion step is weighted by two damping and smoothing parameters. The derived solution after each inversion step should be closer to the minimum and should lie in the area of linearity. Then the next inversion step begins: ray paths and traveltimes are calculated with the updated model.

The misfit between observed and calculated traveltimes has to be minimized with a δm that suites the equation 3.9. We assume a Gaussian error in the relationship: $d_{obs} \approx d_{calc} = G\delta m$. Then a least-square measure of this difference is suitable (eg. Menke (1989)):

$$min d_{obs} - G\delta m^2 \tag{3.10}$$

An objective function $\tau(m)$ is used to weigh the data with their picking errors σ_i, i=1,...,n, which has to be minimized as:

$$\tau(m) = (d_{obs} - G\delta m)^T C^{-1} (d_{obs} - G\delta m), \tag{3.11}$$

where C_d is the covariance matrix with diagonal elements σ_i^2 (eg. Menke (1989)). Equation 3.11 is non-unique when it is under-determined and many δm fit the data. Therefore the least-squares solution is not affected by the unconstrained parameters. A model regularization is introduced to add additional constraints to the inverse problem.

The normalized χ^2 is calculated to estimate the quality of the model fit with:

$$\chi^2 = \frac{1}{N_{res}} \sum_{j=1}^{N_{res}} (\frac{\delta t_j}{\sigma_j})^2, \tag{3.12}$$

where δt_j is the element of d corresponding to the jth traveltime datum, the pick-uncertainty in that datum is labeled with σ_j and the absolute number of traveltime residuals is indicated by N_{res}. If

3.2. INVERSE MODELING WITH 2D TOMOGRAPHY

$\chi^2 = 1$ then a suitable fit is achieved where the model misfit corresponds to the data uncertainty.

3.2.3 Inversion parameterization

Three factors characterize the wide-angle traveltime problem: non-linearity of the traveltime-model relationships, non-uniqueness of the solution and subjective steps during the modeling (Zelt et al., 2003). Non-linearity means that we do not know that the best model (with a global minimum) has been found. To reduce the degree of model non-uniqueness it is possible to use a priori information. We can apply a non-minimum structure model (a detailed starting model), which will usually include more a prior information. But the aim is to take the most objective inverse approach that uses the least subjective information. Therefore we use prior information from forward modeling in a alternating whole model approach with the inversion. The whole model approach was more suitable for the inversion in contradiction to the layer stripping approach. The layer stripping or top to bottom modeling was already applied in the forward modeling step.

The inversion is regularized by smoothing and damping constraints to ensure a stable inversion, preventing the model to contain more structure than the data require. The least subjective portion of a minimum structure model are the first arrivals. This is why later arrivals are not used, because they are buried within the coda. The only exception are reflected phases PmP (oceanic Moho), PtocP (top of the down going plate), PcontP (Moho mantle wedge) to obtain the the corresponding reflector depths. The final tomography solution depends also on the input model. The forward models (input models to the tomography) are updated and refined by the solution of the tomography. This results in an iterative alternating forward and inverse step modeling.

Before applying an inversion to the three wide-angle seismic profiles it was necessary to determine suitable inversion parameters. The parameters should provide satisfying low traveltime residuals (expressed as root-mean square, RMS), a $\chi^2 \approx 1$, velocity model updates after each inversion less than 0.1 % and a low model roughness. For a robust inversion, data outliers are excluded from each iteration by assessing a $\chi^2 \geq 5$. To suppress too large velocity model updates and keep the inversion remain linear, a velocity damping of 25 was applied. This reduces stronger velocity model updates of 6 % from the starting model after the first iteration without damping down to 3 % to stay linear.

A low depth-kernel parameter w results in smaller reflector updates but larger velocity updates. Higher depth-kernel parameters lead to higher reflector updates and smaller velocity updates. For the joint reflection and refraction traveltime tomography, a weighting parameter of $w = 1.0$ and $w = 10.0$ was used.

The Moho reflector inversion strategy follows this approach: with a weighting parameter of $w = 10$ the velocity model is more or less decoupled from the PmP inversions. The picking uncertainties of PmP arrivals are much higher and we do not want to corrupt the velocity model with PmP data (Zelt et al., 2003). In this more objective approach we do not assume that Pn refracts

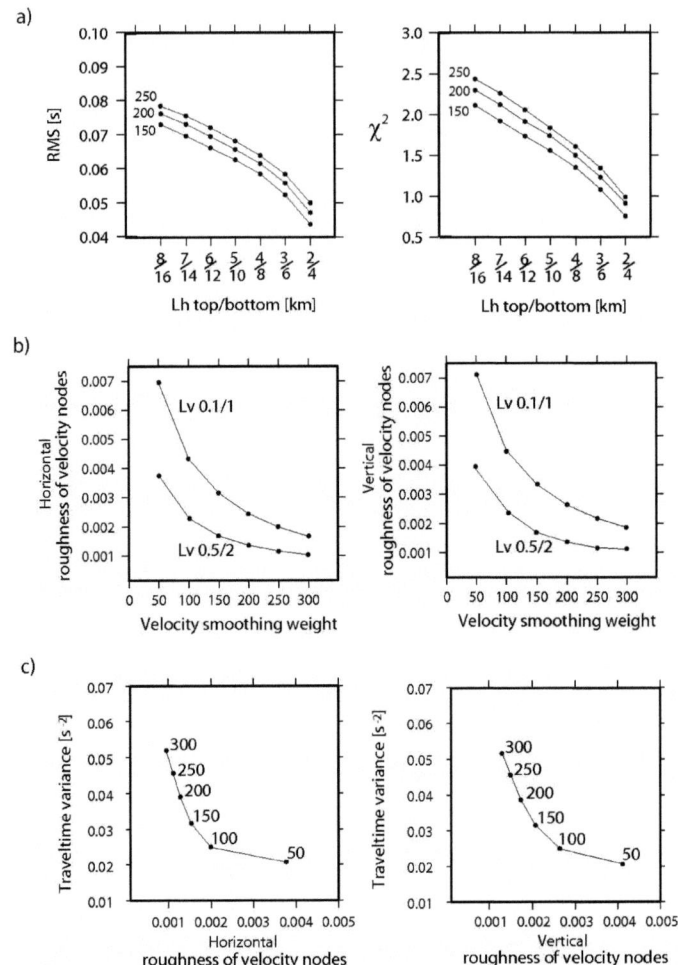

Figure 3.4: Inversion parameter test with profile SO179-P16. a) All data sets with three velocity smoothing factors (150 - 250) showing the dependency between the varying horizontal correlation length Lh (at top and bottom of the model), the RMS value, χ^2 and the model roughness. b) After fixing horizontal correlation length with Lh 3 and 6 km, the vertical correlation length was fixed at 0.5 and 2 km. c) A velocity smoothing of 250 satisfies a traveltime variance and a low model roughness.

3.2. INVERSE MODELING WITH 2D TOMOGRAPHY

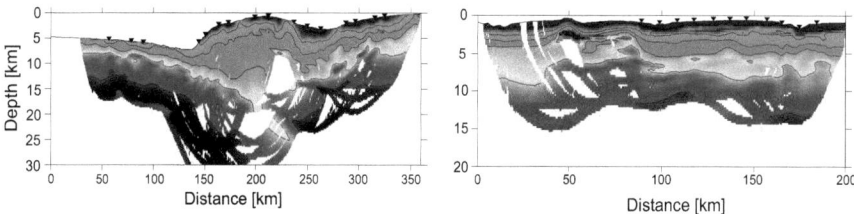

Figure 3.5: Inverted model with horizontal correlation lengths of 2/4 km at the top/bottom of the model, vertical correlation lengths of 0.1/1 km and a velocity smoothing of 200. Smaller correlation lengths results in a loss of model smoothness. The traveltime misfit is low but the models are overfitted.

directly beneath the interface from which PmP reflects. The Moho can be a transition zone, where PmP picks correspond to the top of the zone and Pn refracts at the bottom.

The main critical inversion parameters are: smoothing and correlation lengths to obtain reliable results. Profile SO179-P16 was used to test the dependence between the main critical inversion parameters. The velocity grid is sampled every 0.5 km in the horizontal and every 0.25 km in the vertical direction, which should be fine enough to resolve the expected velocity variations. A single-step inversion approach was applied to the SO179-P16 data set. Each single inversion uses the identical starting model and traveltimes.

Three different sets of inversions were created: in the first test the varying parameter was the horizontal correlation length Lh, which was varied between 2 - 8 km the top and 4 - 16 km at the bottom of the model. The dependence of the velocity smoothing weight was investigated with respectively fixed velocity smoothing parameters with values of 150, 200 and 250. The resulting RMS and χ^2 values show (Fig. 3.4a) that the accuracy is increasing with small horizontal correlation lengths. Furthermore, the accuracy and the χ^2 values are dependent on the velocity smoothing weights. Higher velocity smoothing results in higher RMS and χ^2 values. Therefore a short horizontal correlation length of 3 km at the top and 6 km at the bottom of the model satisfies both RMS and χ^2 values. Lower correlation lengths would overfit the data ($\chi^2 \leq 1$, Fig. 3.5)

In the next test record the dependency between vertical correlation length and the model roughness was investigated. Also the identical starting model and traveltimes were used with a fixed horizontal correlation length Lh of 3 km and 6 km, respectively. Two different inverted datasets were created with varying vertical correlation lengths of 0.1 - 0.5 km at the top and 1 - 2 km at the bottom of the model (Fig 3.4b). Model roughness is the inverse of smoothness and is derived by using the second spatial derivative of the discretized slowness values (Zelt and Barton, 1998). Simultaneously the horizontal and the vertical model roughness increases with decreasing vertical correlation lengths. With increasing velocity smoothing the roughness is getting smaller. Korenaga et al. (2000) stated

30 CHAPTER 3. MODELING THE SEISMIC P-WAVE VELOCITIES FROM WIDE-ANGLE DATA

Model parameter	SO179-P16	SO179-P18	SO179-P19
P phases	10080	13655	6314
PmP phases	466	237	-
PtocP phases	346	593	-
PcontP phases	107	577	-
depth-kernel weight	1.0/10.0	1.0/10.0	0.1
velocity smoothing (wsv)	250	250	250
depth smoothing (wsd)	15	15	15
velocity damping (wdv)	25	25	-
depth damping (wdd)	-	-	-
hor corr lgt top [km]	3.0	3.0	3.0
hor corr lgt bottom [km]	6.0	6.0	8.0
ver corr lgt top [km]	0.5	0.5	0.25
ver corr lgt bottom [km]	2.0	2.0	1.5
forward star	5x10	5x10	5x10

Table 3.1: Inversion parameters applied to the three wide-angle seismic profiles.

a trade-off between correlation lengths and smoothing weights: a higher smoothing weight compensates a lower correlation length. By applying moderate smoothing weight and correlation lengths to the starting model the inversion is prevented to fall in local minima by over determining pre-existing structures (this can be ascribed to non-linearity, Fig. 3.5). A vertical correlation length of 0.5 and 2 km was chosen.

After fixing the correlation lengths a velocity smoothing weight which satisfies a minimum RMS value was chosen in a last parameter testing. The roughness of the model output should be minimized with appropriate velocity smoothing weights but at the same time the data variance S_N^2, where

$$S_N^2 = \frac{1}{N}\sum_{i=1}^{N}(x_i - \bar{x})^2,$$

should be satisfied. A velocity smoothing value of 250 is used for all inversions (Table 3.1). Short-wavelength variations in the observed data require a fine velocity node spacing to obtain a $\chi^2 = 1$. Therefore the gridnode size decreases with depth. For the first arrival traveltime tomography the picks from the raytracing procedure were re-inspected and re-picked for higher accuracy reasons. The pick density was increased, therefore every usable trace was picked. Higher uncertainties in greater offsets often induced a lower pick density of every second or at least every third trace. On profile SO179-P16 traveltimes were picked from 13 stations.

Profile SO179-P18 uses data from 20 stations, which are 20 % more picks compared to SO179-P16 and thus results in a higher raycoverage. Model dimensions of both dipping profiles are equal (344-360 km x 40km), therefore the same model parameters are used. SO179-P19 provide traveltimes on 11 stations for the tomography with slightly different parameters (Table 3.1).

3.3 Results of the forward and the inverse modeling

3.3.1 Testing different input models for the tomography

In this chapter a combined analysis of the forward and inverse models is presented.
The eastern dip line SO179-P16 resolves tectonic features which merit a detailed investigation: the steep trench slope, the isolated forearc high, the uplift motion and the trench retreat (refer to chapter 1.1). These are evidences for a formerly subducted seamount. For this purpose the velocities in the region where the seamount is expected are systematically changed in the transition zone to the backstop. The velocities are changed in a 50 x 5 km wide region in 10 - 15 km depth and from profile-km 150 - 200 (Fig. 3.6). The inversion was calculated twice with two input models: one with higher velocities (5.8 km/s), as expected for a seamount, and one model with lower velocities (5.2 km/s). The determined model differences between the forward and the inverted models (Fig. 3.6) presents a pattern of 0.1 km/s isolines with general low model differences. In the test region of 50x5 km the inversion tries to fit the input velocities to the real data. The forward model velocities are too high by 0.4 km/s in the model with the suggested seamount. The alternative input model provides too low velocities which are increased by 0.2 km/s to fit the data. The inverted models provide some evidence for a subducted seamount: the velocities of the active outer wedge merge smoothly into the inner wedge. The presence of a backstop with a sharp velocity contrast and a high velocity gradient could not be validated. However, the data set supports the existence of a subducted large scale body.

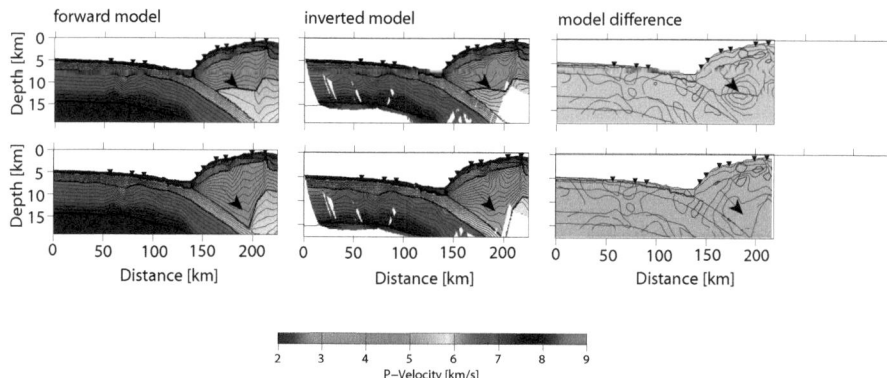

Figure 3.6: Two different input models to the tomography testing the hypothesis for a subducted seamount. The left panel provide the input (forward-) model with increased (top) and decreased (bottom) velocities, pointed by the arrow. The inverted model (middle panel) and the model difference (right panel) display evidences for higher velocities in the forearc high, close to the backstop.

32 CHAPTER 3. MODELING THE SEISMIC P-WAVE VELOCITIES FROM WIDE-ANGLE DATA

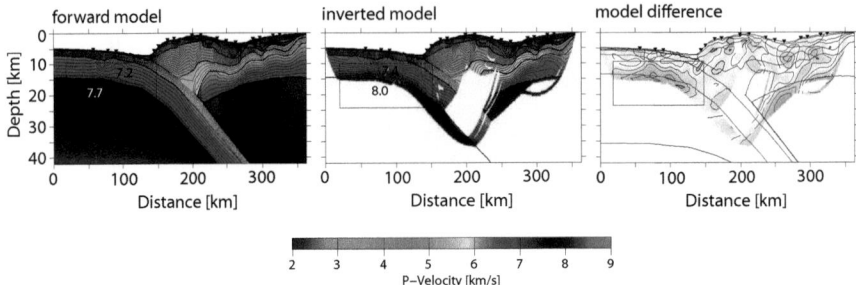

Figure 3.7: Testing lower mantle velocities on profile SO179-P16. The forward model provides reduced lower velocities in lower crust with 7.2 km/s and in the mantle with 7.7 km/s. The inverted model applies higher velocities to the mantle (8 km/s) and to the lower crust (7.4 km/s). The model difference (right panel) demonstrates the increased velocities (positive isolines) after the inversion by red colors.

Another test examines the oceanic mantle velocities and the lower oceanic crustal velocities. The western profile SO179-P18 yields low mantle velocities of 7.7 km/s. Therefore an input model for profile SO179-P16 with comparable low mantle velocities (Fig. 3.7) was prepared for the inversion. The velocities of 7.4 km/s above the crust-mantle boundary are 0.2 km/s higher compared to the western profile. For this reason the velocities where reduced in the forward model to 7.2 km/s. A joint inversion of reflected PmP phases and refracted phases was applied with a depth weighting kernel of $w = 1$, where reflector and velocity updates are equal.
The inverted model and the model difference plot (Fig. 3.7) show an increase of the mantle velocities and an increase of the lower crustal velocities, which reinforce the higher mantle velocities on profile SO179-P16.

3.3.2 Final forward and inverse P-wave velocity models

In this section the final forward and inverse velocity-depth models are presented individually. The forward model results of the dipping profiles SO179-P16/P18 are displayed in Figure 3.8. These models are the result of the alternatively executed forward and inverse process to investigate the velocity-depth structure. The forward models are based on a top to bottom approach (refer to chapter 3.1), and obtain a maximum depth of 40 km. The unresolved regions are highlighted in light gray. The two profiles are aligned along the trench (Fig. 3.8, dashed line).
The inversion was weighted with a depth-kernel of $w = 1$, which determines equal velocity and reflector updates. Every tomographic model is based on a joint inversion of reflected and refracted phases. Each model was calculated with four iterations. The inverted models expose a poor raycoverage (white patches, where less than 5 rays travel through the grid) in the central portion of the model (Figs.

3.3. RESULTS OF THE FORWARD AND THE INVERSE MODELING

3.9 and 3.10). A strong velocity gradient and high velocities force the rays to deeply penetrate the model. One stopping criterion was velocity-model updates less than 0.1 %. The RMS value of 53 - 63 ms was reached, which represents a good result and is an compromise between model resolution and traveltime residuals.

The derivative weight sum in Figs. 3.9 and 3.10 is the column-sum vector of the normalized Fréchet derivative matrix. It is the weighted sum of the path lengths influenced by certain model parameters. DWS value for a grid cell does not only account for a number of cells hits per ray, it is rather dependent on ray length in the cell and pick-uncertainty of each ray. Therefore the DWS vector is an assessment of solution sensibility based on data quality and quantity. High DWS values can be understood either as a denser sampling of rays or as an accumulation of higher quality ray paths with smaller pick-uncertainties. White areas correspond to grid cells with DWS values less than 5 (Figs. 3.10 and 3.9). The model differences between forward and inverse model are quite small with +0.1/-0.2 (SO179-P16) and +0.2/0.2 km/s (SO179-P16).

34 CHAPTER 3. MODELING THE SEISMIC P-WAVE VELOCITIES FROM WIDE-ANGLE DATA

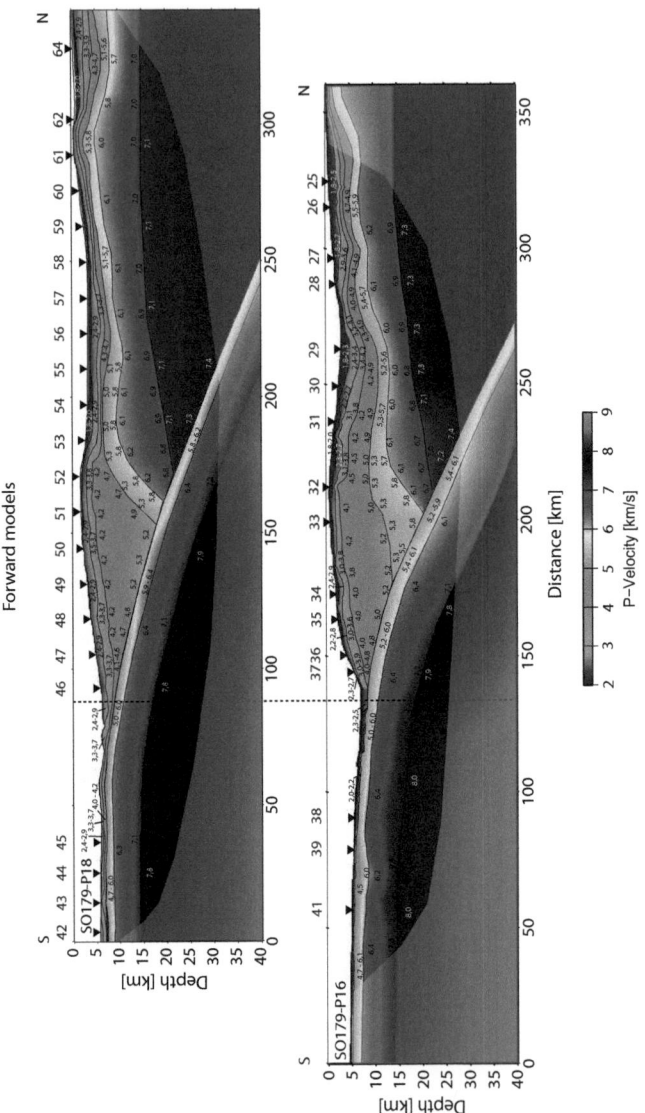

Figure 3.8: Forward models of profiles SO179-P18 (top) and SO179-P16 (bottom). The P-wave velocities are color coded and additionally marked by numbers [km/s]. Solid lines indicate layer interfaces. Station locations are indicated by triangles. The grey shaded area is not well resolved. The models are aligned to the trench (dashed line).

3.3. RESULTS OF THE FORWARD AND THE INVERSE MODELING

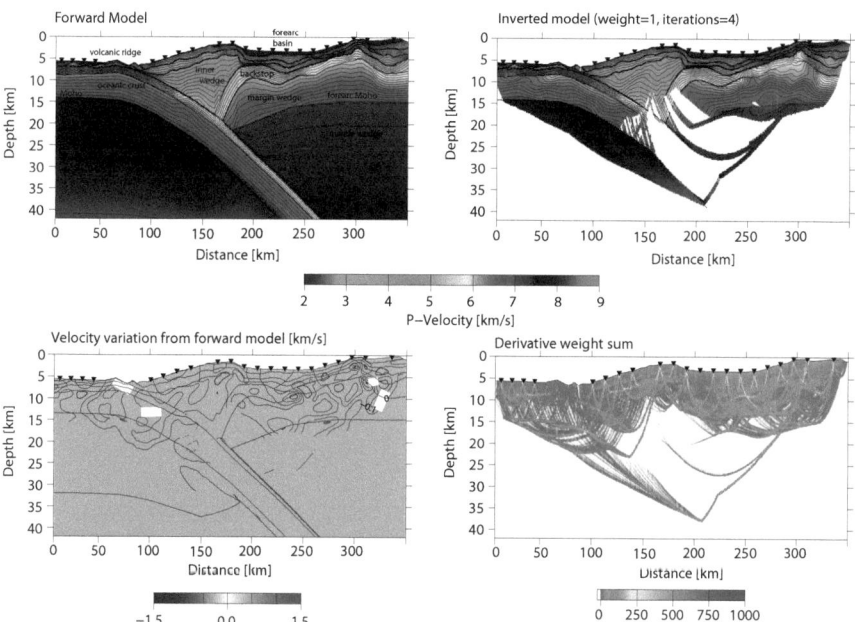

Figure 3.9: Final inverted model of profile SO179-P18. The upper left figure displays the input model for the inversion, which corresponds to the forward model based on MacRay. The main tectonic features are labeled. The upper right presents the inversion result based on the Tomo 2D code by (Korenaga et al., 2000) after four iterations. The white patches are poorly resolved regions with less than five rays per cell. The lower left figure displays the difference between the input forward and resulting inverted model. The maximum velocity variation iso-lines are ±0.2 km/s. The lower right figure shows the color coded DWS matrix, which reflects the ray density and the ray path quality.

36 CHAPTER 3. MODELING THE SEISMIC P-WAVE VELOCITIES FROM WIDE-ANGLE DATA

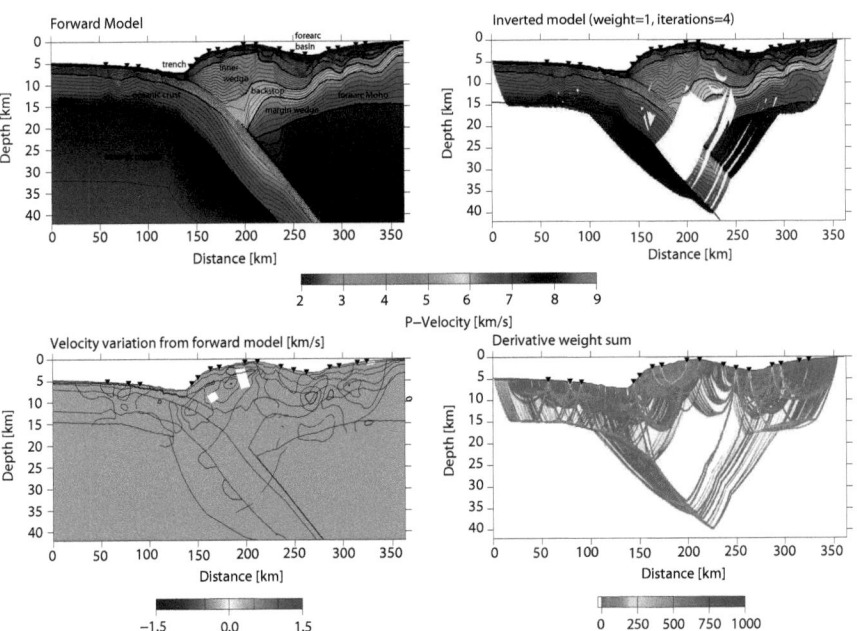

Figure 3.10: Final inverted model of profile SO179-P16. For a detailed description please refer to the previous Fig. 3.9

3.3. RESULTS OF THE FORWARD AND THE INVERSE MODELING

3.3.3 Ocean basin, trench and subducted plate

The incoming plate off central Java is covered by approximately 500 m of pelagic/hemipelagic sediment (Layer 1). Bending related faulting is observed in the mini streamer seismic section (Fig. 3.11). The distance between the faults is about 2 - 10 km, with a fault length of 5 - 20 km (Kopp et al., 2006). Profile SO179-P18 provides a 2 - 3 times thicker layer of sediments and volcanic material with a lower velocity gradient compared to the eastern profile (Fig. 3.8). The trench sediment fill on the eastern seismic reflection profile has a thickness of $\approx$ 500 ms TWT, which corresponds to hundred meters sediment thickness. The trench sediments are presumably originated from the upper plate caused by the steep slope angle resulting in gravitational mass wasting. Bending related faults penetrate the sediment coverage (Fig. 3.11).

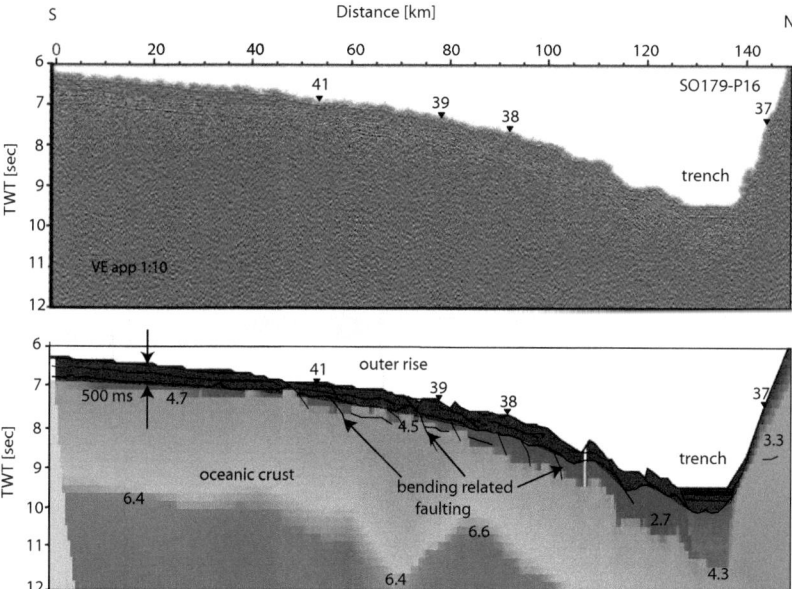

Figure 3.11: Water migrated seismic record section of profile SO179-P16 (top) and interpretive line drawing (bottom) superimposed with a velocity-depth model of the inversion. Velocity values are noted as numbers. This section only shows the outer rise. Station locations are indicated by triangles.

38 CHAPTER 3. MODELING THE SEISMIC P-WAVE VELOCITIES FROM WIDE-ANGLE DATA

3.3.3.1 Obtaining Moho depth by joint inversion of refracted and reflected traveltimes

The quality of PmP phases was sufficient to pick 466 PmP arrivals on profile SO179-P16 and 237 arrivals on profile SO179-P18, respectively (Tab. 3.1), to run an joint inversion with reflected and refracted rays. The eastern profile provides more PmP arrivals and a better ray coverage of oceanic mantle reflections down to 25 km bsf from stations on the forearc high close to the trench. Therefore a squeezing test for the Moho depth was applied to this profile. A joint reflection- and refraction tomography with PmP phases and all refracted phases was applied to profile SO179-P16. Different Moho depths and curvatures were used to check the reliability of the results (Fig. 3.12). The same inversion parameters were used for each iteration and a depth-kernel weighting parameter of $w = 10$ was applied, which determines larger reflector updates and only small velocity updates to the inversion. The reflector was sampled with reflector nodes every kilometer.

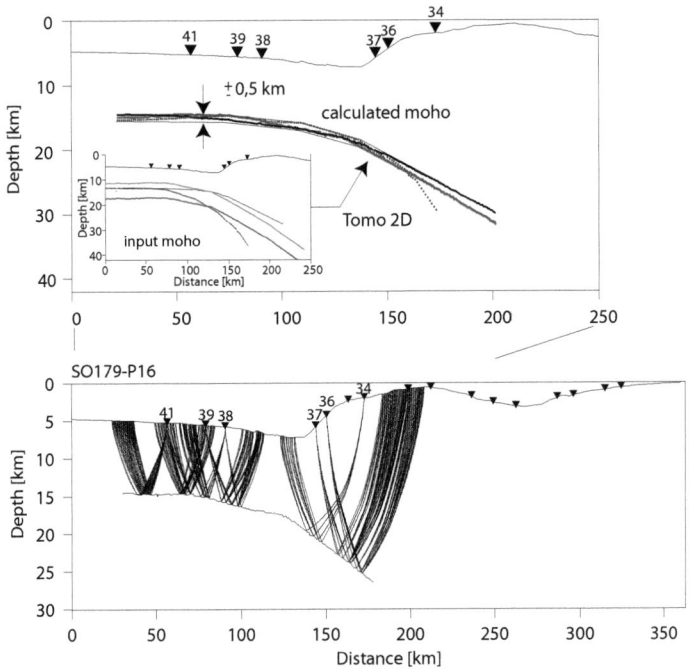

Figure 3.12: Top: Moho reflectors for the joint inversion. All models were inverted with the same starting model and same parameters. The calculated Moho reflectors converge in a depth range of 15 ± 0.5 km. Bottom: ray coverage of all reflected PmP phases. The area between stations 38 and 37 is not resolved and the Moho is interpolated.

3.3. RESULTS OF THE FORWARD AND THE INVERSE MODELING

The derived Moho depths (Fig. 3.12) converge in a depth interval of 1 km at a mean Moho depth of 15 km, providing a maximum uncertainty of ± 0.5 km. The Moho interface between profile-km 100 and 150 is interpolated due to a data gap of PmP arrivals since no station could be deployed in the trench because of the great water depth here. Nevertheless, the Moho depth from the forward modeling could be reinforced by the tomographic inversion.

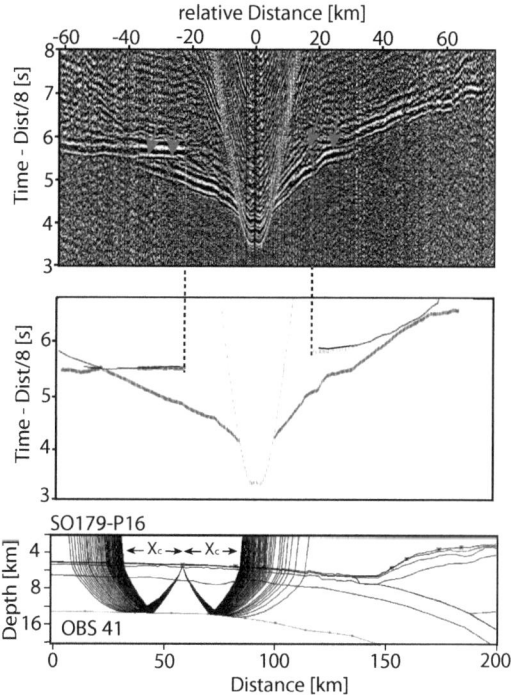

Figure 3.13: Critical wide-angle reflections to constrain the Moho depth with stronger amplitudes at offsets greater than the critical distance Xc. Red arrows indicate strong amplitudes of PmP reflections.

At the critical angle of incidence the ray energy is partitioned into a wide-angle reflected and a refracted head wave that travels along the interface. The amplitudes of refracted arrivals as a function of the offset are generally smaller than amplitudes of wide-angle reflections at offsets greater than the critical distance Xc. For this reason an additional constraint for the Moho depth is the critical wide-angle reflection. At this certain critical distance Xc the energy of the refracted ray does not penetrate into the mantle and the corresponding headwave travels along the Moho (Stein and Wysession,

2003). Figure 3.13 shows the traveltime fit of increased amplitudes at offsets greater than the critical distance. The PmP reflections are clearly evident and the onset of strong amplitudes fits with the forward model critical rays (dashed line).

As a result the Moho depth of the eastern profile SO179-P16 is constrained by applying different input Moho interfaces (shape and depth), which are converging in a distinct depth interval of ± 0.5 km and increased amplitudes of wide-angle reflections at offsets greater than the critical angle.

The Moho depth of SO179-P18 is constrained only seaward of the trench due to the lack of PmP arrivals further landward. The change from oceanic crust to the mantle could be verified with a 1 km thinner oceanic crust compared to the profile SO179-P16 (Fig. 3.8).

42 CHAPTER 3. MODELING THE SEISMIC P-WAVE VELOCITIES FROM WIDE-ANGLE DATA

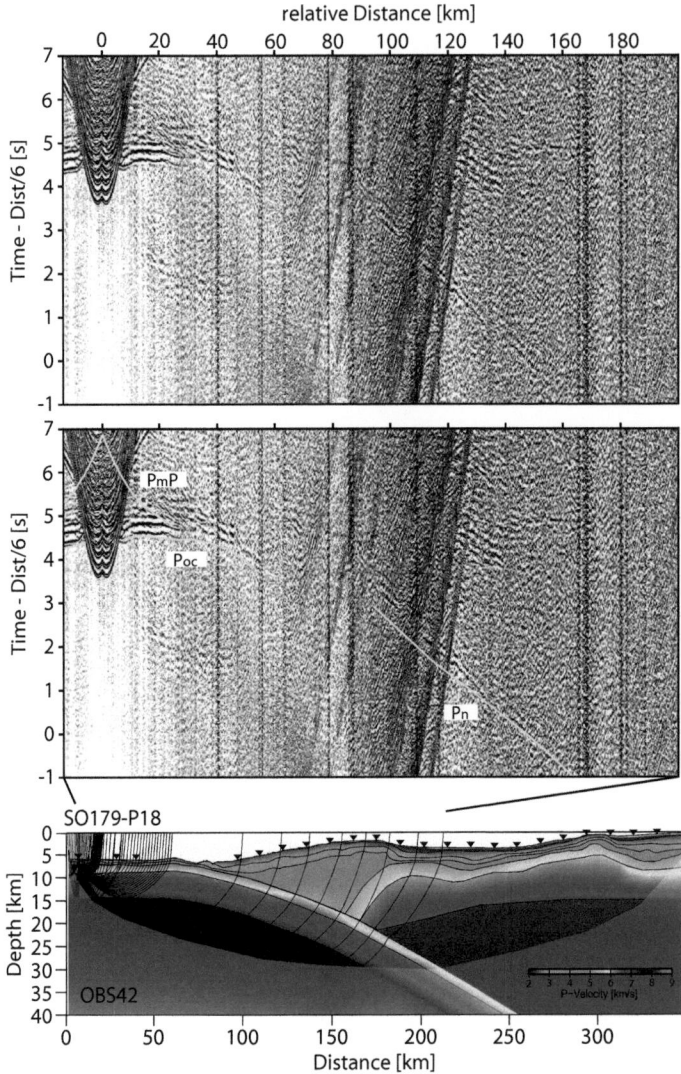

Figure 3.14: Forward result of OBS42: Wide-angle seismic record section of station OBS42 (profile SO179-P18). Calculated traveltimes are in yellow (middle) with corresponding raypaths (bottom).

3.3. RESULTS OF THE FORWARD AND THE INVERSE MODELING

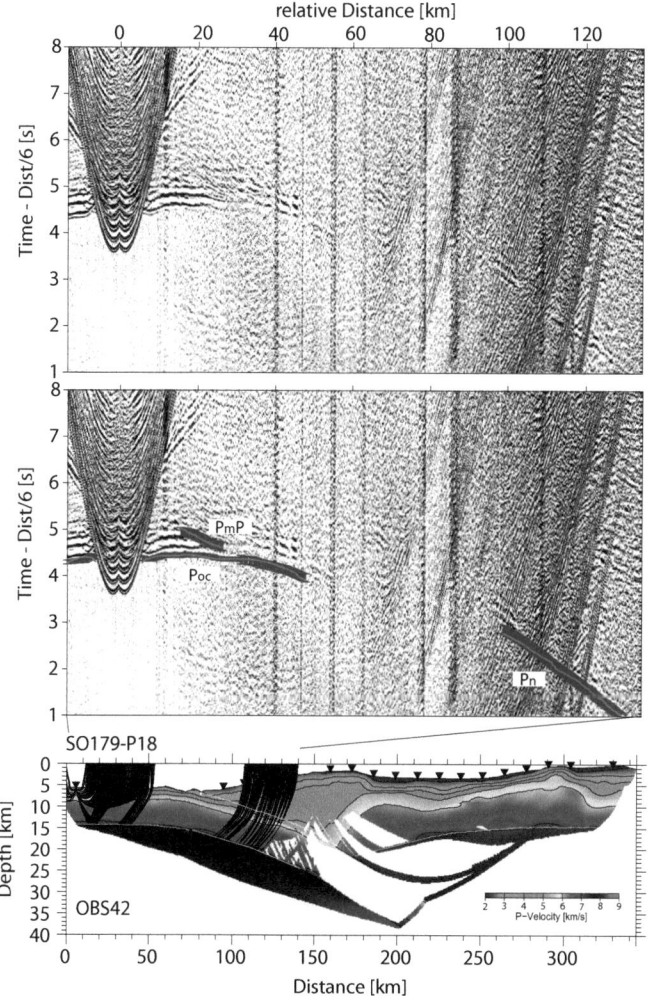

Figure 3.15: Inverted model of OBS42: Wide-angle seismic record sections of profile SO179-P18 (reduced with 6 km/s). Top: interpreted seismic phases, Center: Calculated (red) and picked traveltimes (blue) with error bars, Bottom: Corresponding ray paths through the model for profile SO179-P18.

44 CHAPTER 3. MODELING THE SEISMIC P-WAVE VELOCITIES FROM WIDE-ANGLE DATA

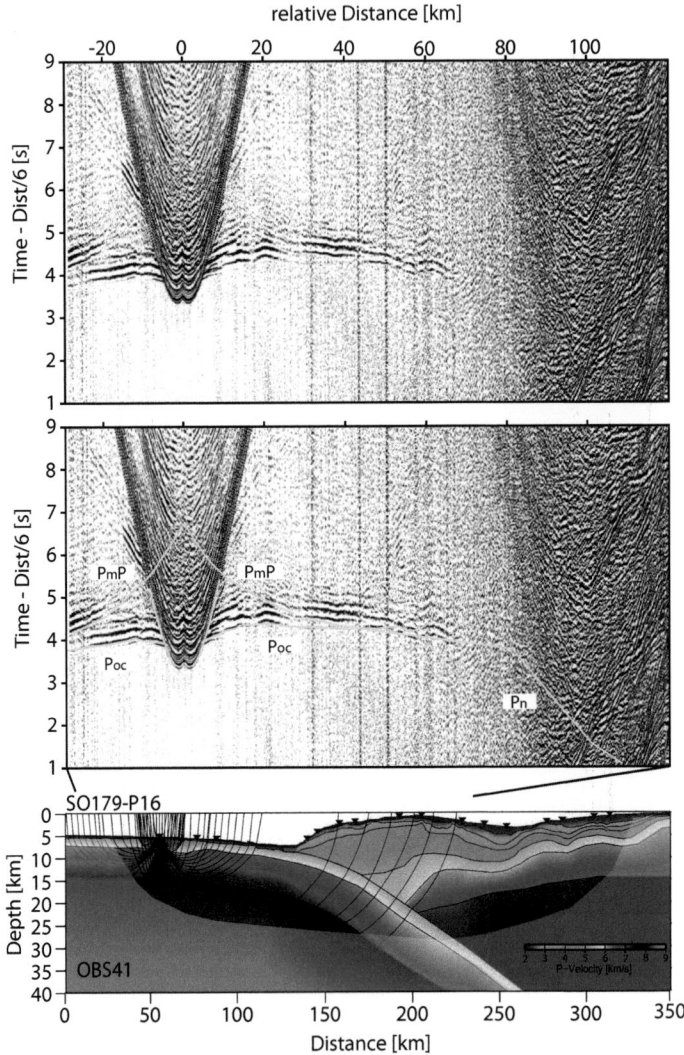

Figure 3.16: Forward model of OBS41, profile SO179–P16. Please refer to Fig. 3.14 for details

3.3. RESULTS OF THE FORWARD AND THE INVERSE MODELING

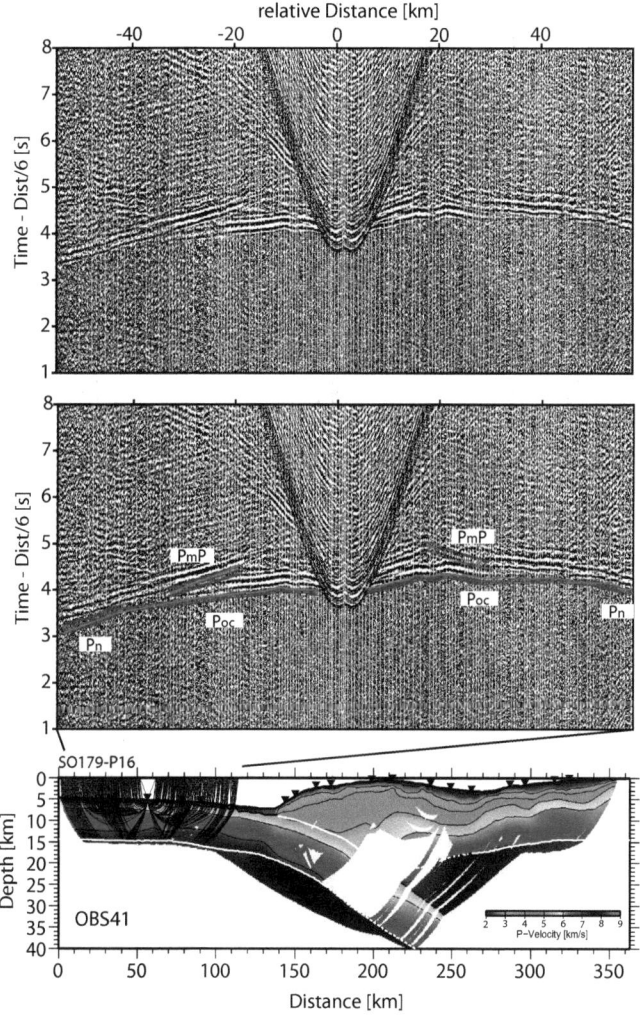

Figure 3.17: Inverted model of OBS41:Wide-angle seismic record sections of profile SO179-P16 (reduced with 6 km/s). Top: interpreted seismic phases, Center: Calculated (red) and picked traveltimes (blue) with error bars, Bottom: Corresponding ray paths through the model for profile SO179-P16.

A velocity of 4.7 - 6.0 km/s was used to model the oceanic layer 2 with a thickness of 3 km based on Poc-phases (Fig. 3.8). In this region, the seismic velocities increase rapidly with depth (gradients of 1 - 2 km/s per kilometer of depth). The upper oceanic crust is composed of basaltic pillow lavas and lava debris in varying degrees of alteration (White et al., 1992). The velocities at the top in this layer increase in the direction of subduction from 5.0 km/s near the trench up to 5.9 km/s at profile-km 135 on profile SO179-P18 and 5.6 km/s at profile-km 225 on profile SO179-P16. The western profile accommodates a local high velocity anomaly in this layer at profile-km 130 (Fig. 3.8).

The lower oceanic crust (Layer 3), displays a velocity increase from 6.3 - 7.1 km/s on profile SO179-P18 and 6.2 - 7.4 km/s on the neighboring profile. The P-wave velocities at the bottom of layer 3 are 0.2 km/s higher compared to SO179-P18. This could be validated by testing the input model with lower velocities in the forward model (Fig. 3.7). Layer 3 of profile SO179-P16 has a thickness of 6 km (at profile-km 65) to 8 km and is mainly gabbroic in composition. These typical oceanic crustal velocity values agree with the southern Sumatra and western Java data (Kopp et al., 2002). PmP reflections yield a Moho depth of 10 km below seafloor seaward of the trench (Fig. 3.8). Kopp et al. (2002) observed a crustal thickness of 7.4 km offshore western Java at 106°E. In this study, a crustal thickness of 10 km on the eastern profile SO179-16 and 9 km on the western profile SO179-18 is determined. The Moho marks the transition from basaltic-gabbroic crust to the peridotitic mantle. The Pn mantle refractions show mantle velocities of 7.7 - 7.8 km/s as can be seen on station OBS42 (Fig. 3.15), whereas OBS41 on the eastern profile provides mantle velocities of 7.9 - 8.0 km/s (Fig. 3.17), which is regarded as 'normal' mantle velocities (White et al., 1992). The raytracing model predicts mantle phases over a distance of 180 km (Fig. 3.14), however the Pn phase can not be traced over the entire distance in the wide-angle seismic section. The mantle phases of profile SO179-P16 do not penetrate deeply into the oceanic mantle due to their limited offsets. The obtained velocities agree with southern Sumatra and western Java transects mantle velocities (Kopp et al., 2002).

3.3.4 Outer- and inner wedge, backstop

The deformation front is located at profile-km 90 on profile SO179-18 and at profile-km 130 on the eastern profile SO179-16. The forearc high at profile SO179-16 is as shallow as 1000 m. Normal faulting (Fig. 3.18) is an indication of a strong uplift compared to the adjacent forearc high (Kopp et al., 2006). The uplift leads to slope failure at the trench wall and hence to the occurrence of slope deposits and subsequent sediment pond structures (Kopp et al., 2006).

The velocity structure of the outer wedge was determined from refracted waves mainly from seven OBH/S stations on the western profile (OBH46 - OBH52) and six stations on the eastern profile (OBH37 - OBH32) located on the frontal prism and forearc high. On profile SO179-18 an approximately 1600 m thick layer of unconsolidated sediment with velocities of 2.4 - 2.9 km/s is superimposed by a second layer with approximately the same thickness and velocities of 3.3 - 3.7 km/s (Fig. 3.8). This determines upper sedimentary units with a total thickness of 3.5 km. These layer velocities

3.3. RESULTS OF THE FORWARD AND THE INVERSE MODELING

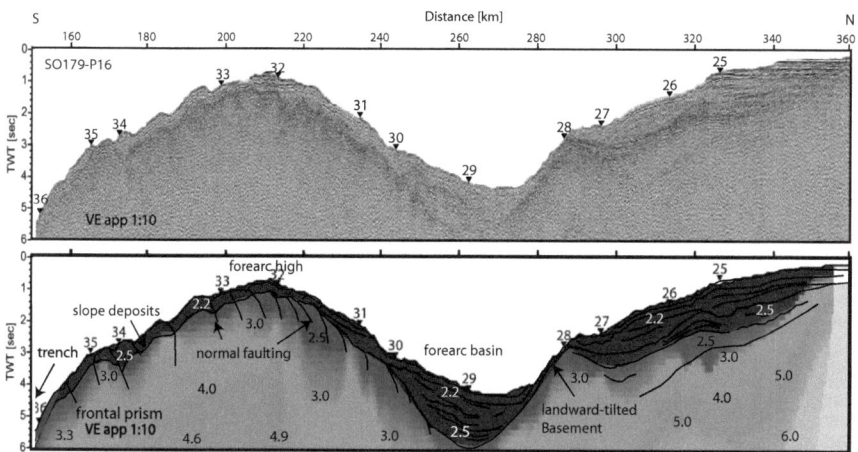

Figure 3.18: Water migrated seismic record section of profile SO179-P16 (top) and interpretive line drawing (bottom). This section is the landward continuation of Figure 3.11. Station locations are indicated by triangles. The tomographic velocity-depth model is superimposed and the seismic P-wave velocities [km/s] are notified as numbers.

remain constant (undisturbed) over the entire outer wedge whereas the thickness decreases slightly to 2.6 km at the crest of the forearc high.

48 CHAPTER 3. MODELING THE SEISMIC P-WAVE VELOCITIES FROM WIDE-ANGLE DATA

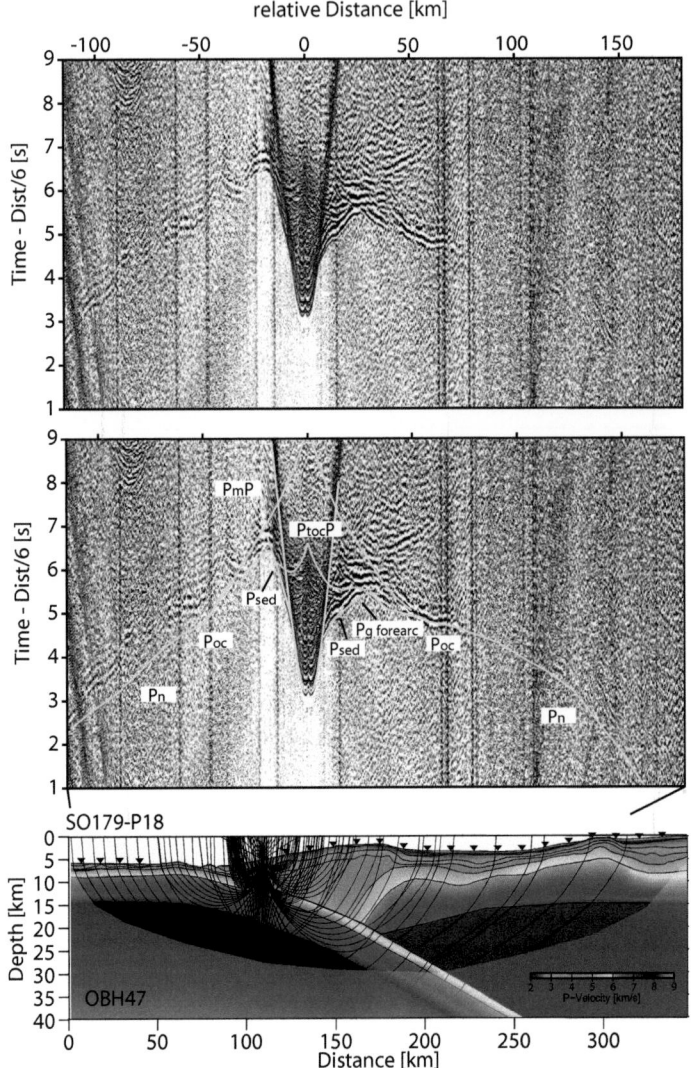

Figure 3.19: Forward model of OBH47, profile SO179-P18.

3.3. RESULTS OF THE FORWARD AND THE INVERSE MODELING

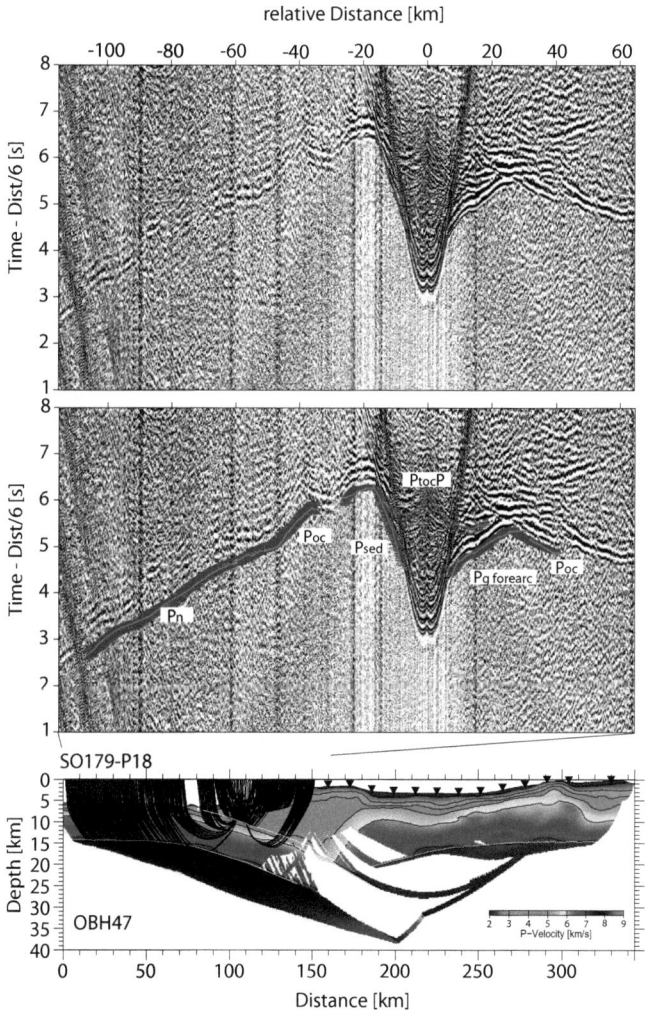

Figure 3.20: Inverted model of OBH47.

50 CHAPTER 3. MODELING THE SEISMIC P-WAVE VELOCITIES FROM WIDE-ANGLE DATA

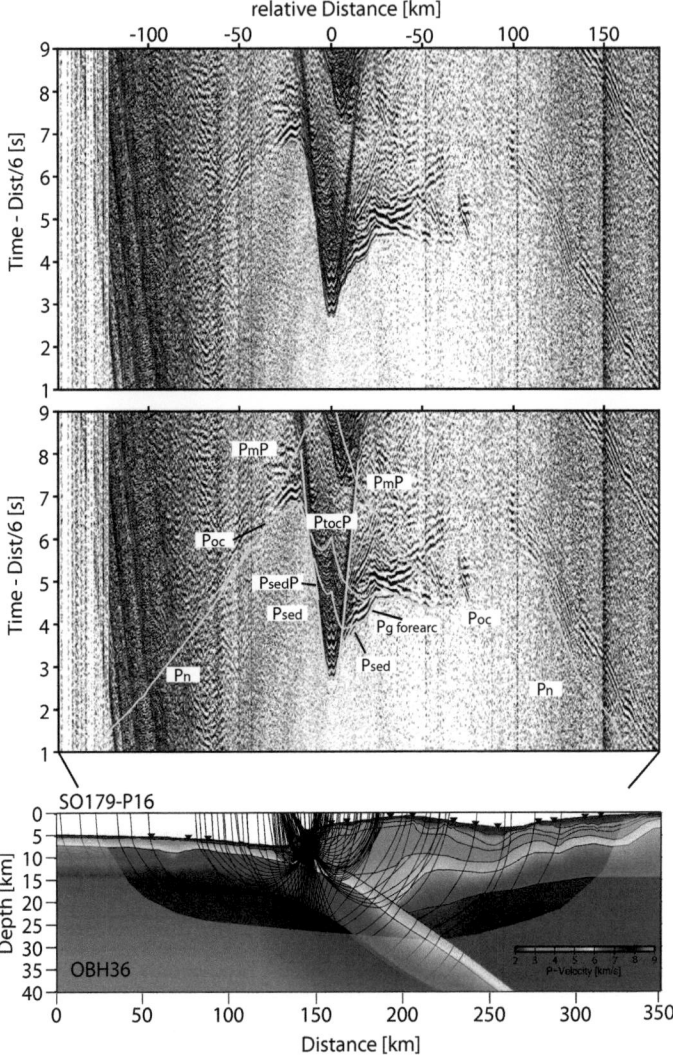

Figure 3.21: Forward model of OBH36, profile SO179-P16.

3.3. RESULTS OF THE FORWARD AND THE INVERSE MODELING

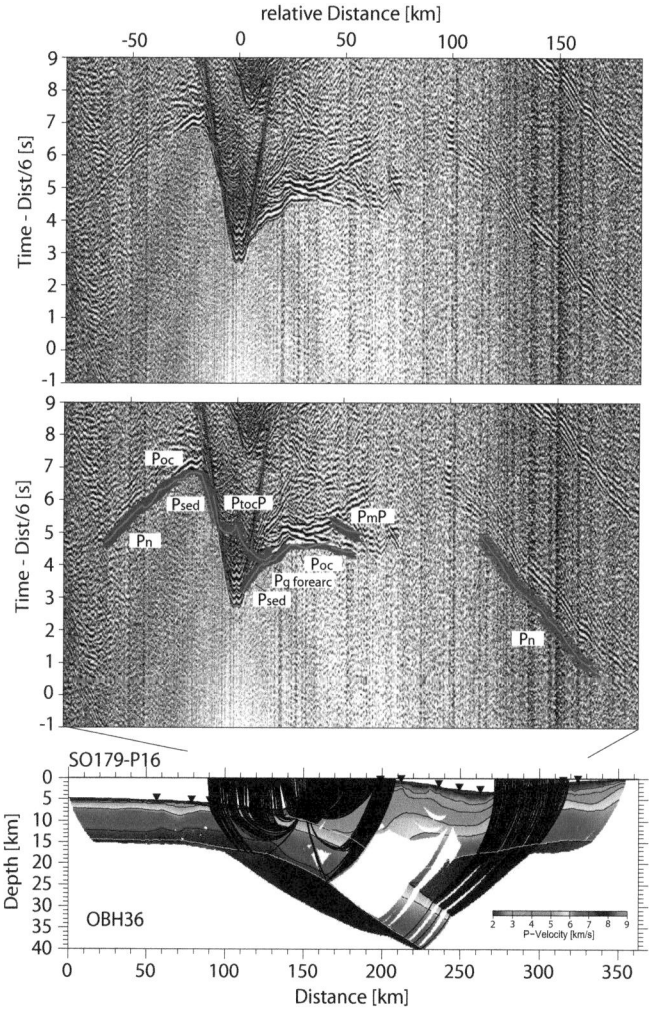

Figure 3.22: Inverted Model of OBH36.

52 CHAPTER 3. MODELING THE SEISMIC P-WAVE VELOCITIES FROM WIDE-ANGLE DATA

The outer wedge below the sedimentary units displays velocities of 4.2 km/s and 5.2 km/s just above the underthrusting plate. At profile-km 150-170 the inner wedge impinges the backstop to the margin wedge. The velocity contrast of 0.4 - 0.6 km/s defines a sharp contrast between the inner and margin wedge. The lowest point of the crustal forearc unit is positioned below the crest in 20 km below sea level at profile-km 150 with a P-wave velocity of 6.8 km/s (Fig. 3.8).
The sedimentary units on profile SO179-16 vary with different layer thicknesses over the entire distance between the trench and the crest of the forearc high. The thinnest uppermost sedimentary layer displays low velocities of 1.8 - 2.0 km/s turns up at the slope of the forearc basin at profile-km 150. The thickness remains constant, until it merges into the forearc basin, were it increases in thickness to approximately 2 km and where the bottom P-velocity increases to 2.3 km/s. The seismic section reveals a highly faulted forearc high, where water penetrates the outer wedge (Fig. 3.18). The sediment thickness increases along the rising forearc slope, which is determined by slope deposits, trapped by fault escarpments e.g. at OBH34 (Fig. 3.18). Beneath the uppermost, unconsolidated sedimentary layer follows a few hundred meters thin layer, which remains constant in thickness, until it merges into the forearc basin. The P-wave velocities range from 2.3 - 2.7 km/s at the trench axis to 2.4 - 2.9 km/s. The lowest and most consolidated sedimentary unit displays velocities of 3.0 - 3.9 km/s on profile SO179-16. The total thickness of sediments is approximately 1 km at the trench and up to 4 km at profile-km 225, just behind the crest of the forearc high.
Beneath the sedimentary units the velocities gradually increase from 4.0 - 5.2 km/s, flanking the top of the underthrusting oceanic crust. In contrast to profile SO179-P18 the boundary between the inner and margin wedge is aligned subhorizontal at the deepest part, with a low velocity contrast of 0.1 (at profile-km 175) to 0.6 km/s (at profile-km 200). At the crest of the forearc high it reaches the maximum depth and continiues in a constant depth of 10 km until the forearc basin.
PtocP reflections from the the top of the oceanic crust could be modeled with station OBH47 and OBH36 (Fig.s 3.19 and 3.21). The downgoing oceanic crust subducts at an angle of approximately 10° below the margin wedge.

3.3.5 Forearc basin, margin wedge

The forearc basin on SO179-P18 is 90 km wide and extends along the profile in NS direction from profile-km 180 (OBH53) to the shelf at profile-km 260 (OBH59). The layers appear mainly undeformed and reach a thickness of 4 km. The seismic velocities in the basin range from 1.8 km/s at the top to 3.7 km/s at the base of the basin, which are highly compacted and lithified sediments with a volcaniclastic origin.
The convex lens shaped boundary of the inner wedge is divided into two parts: a 300 - 500 m thin layer with velocities from 5.3 - 5.8 km/s. The second layer with velocities of 6.1 - 6.9 km/s and a maximum layer thickness of 10 km between stations OBH53 and OBH54 verging the mantle boundary at the bottom. The layer is thinning out to OBS56 (Fig. 3.8). The thinnest part is located at OBS56

3.3. RESULTS OF THE FORWARD AND THE INVERSE MODELING

with an increasing thickness beyond this station in landward direction.

The deepest point of the forearc crust at 20 km bsf coincides with the crest of the forearc high at profile-km 170 km. The mantle boundary alignment is slightly flattening from the deepest point into landward direction to 13 km depth bsf at OBS57, which could be verified by 577 PcontP phases on this profile (Tab. 3.1). OBS53 and OBH58 displays the fit of PcontP phases (Fig's 3.23, 3.24, 3.25 and 3.24). The mantle boundary remains in a constant depth at 15 km.

The forearc basin on SO179-P16 is 50 km wide (OBH31 - OBS28) and approximately 5 km deep (Fig. 3.8). The basin strata onlap at the outer forearc high and are tilted landward (Fig. 3.18). The layers are tilted and disturbed with its deepest point (OBS29) in the middle of the basin and the thinnest portions at the rim. The velocity gradients in the forearc basin are much lower compared to the western profile. The P-wave velocities at the base of the sediments are 3.8 - 4.2 km/s.

The main difference to the profile SO179-P18 is the boundary to the margin wedge: there is no sharp boundary, the velocity contrast is much smaller (0.1 - 0.3 km/s) in seaward front of the boundary to the inner wedge, compared to 0.4 - 0.6 km/s. The boundary alignment is subhorizontal, especially in the deeper portion of the forearc, close to the downthrusting plate. The P-wave velocities in the inner wedge are comparable to SO179-P18 with 5.3 - 5.8 km/s(Fig. 3.8). Compared to the western profile the deepest point of the forearc crust is positioned in the same depth of 20 km depth, located below the forearc crest. The crust mantle boundary is slightly flattening from the deepest point into landward direction to 15 km bsf depth. The crust mantle boundary could be verified by only 20% of PcontP phases.

54 CHAPTER 3. MODELING THE SEISMIC P-WAVE VELOCITIES FROM WIDE-ANGLE DATA

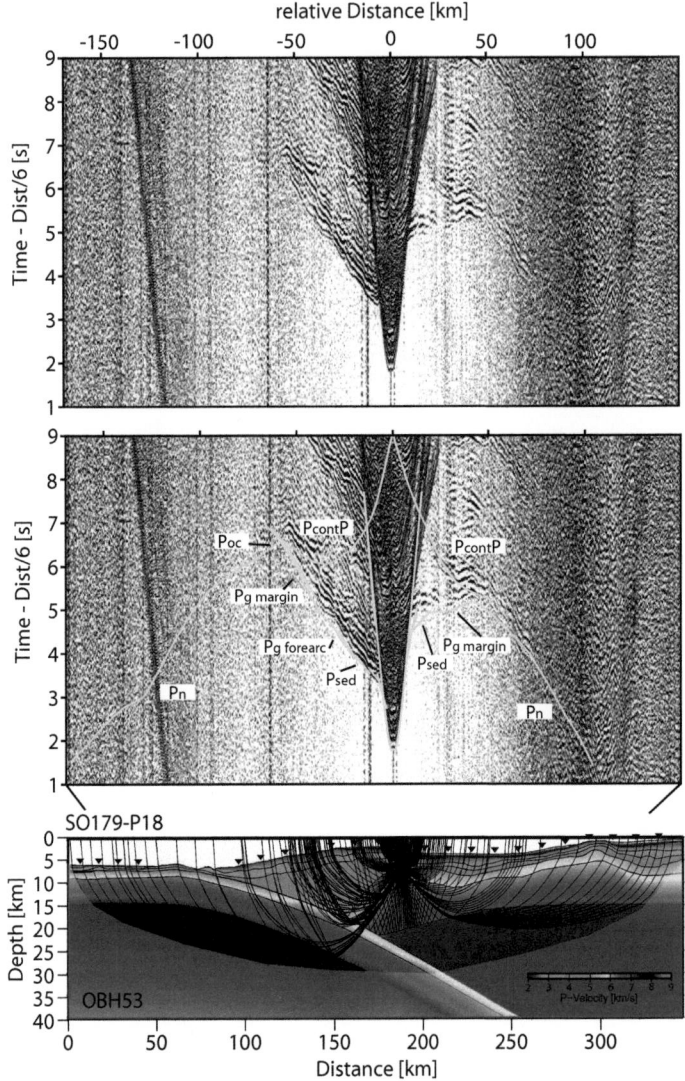

Figure 3.23: Forward model of OBH53, profile SO179-P18.

3.3. RESULTS OF THE FORWARD AND THE INVERSE MODELING

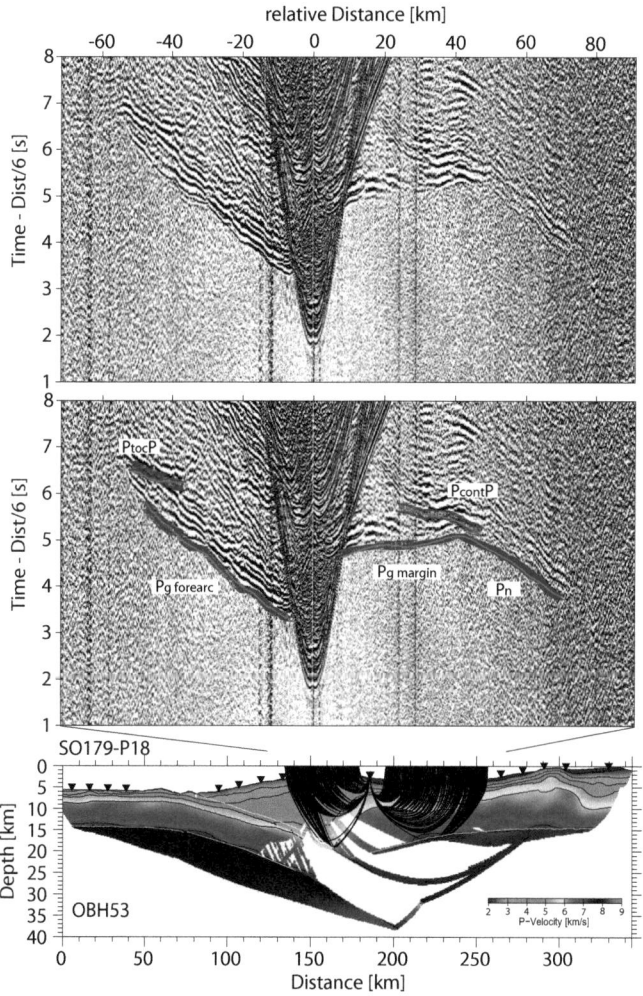

Figure 3.24: Inverted model of OBH53.

56 CHAPTER 3. MODELING THE SEISMIC P-WAVE VELOCITIES FROM WIDE-ANGLE DATA

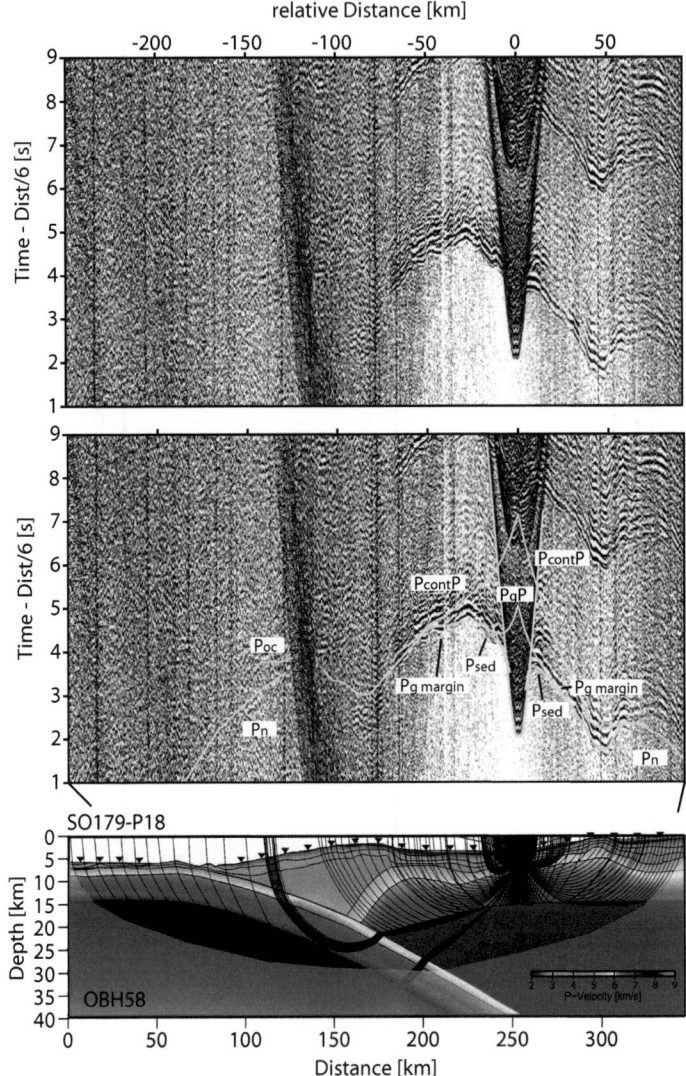

Figure 3.25: Forward model of OBH58, profile SO179-P18.

3.3. RESULTS OF THE FORWARD AND THE INVERSE MODELING

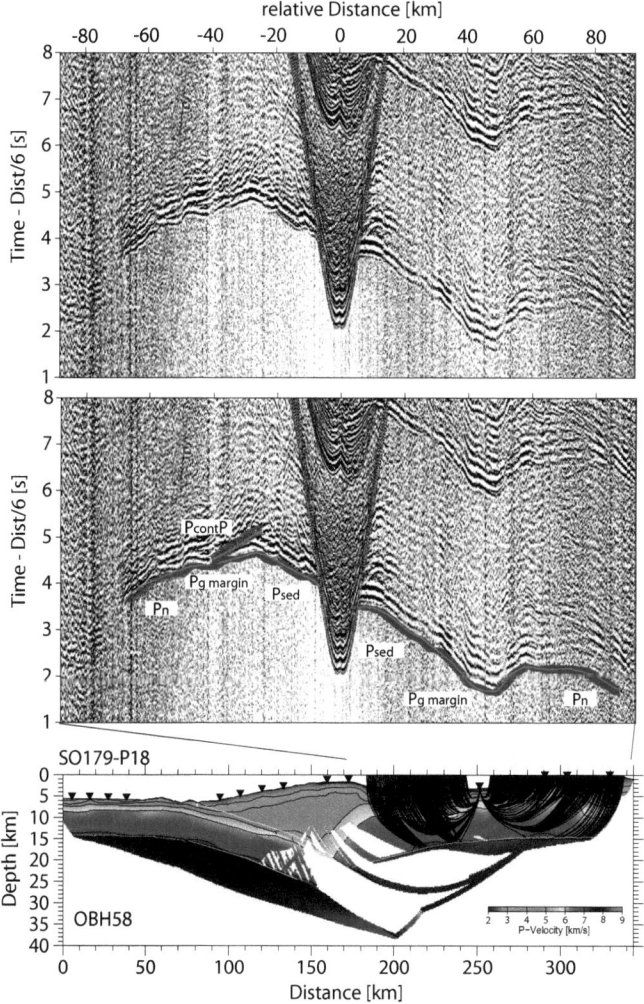

Figure 3.26: Inverted model of OBH58

58 CHAPTER 3. MODELING THE SEISMIC P-WAVE VELOCITIES FROM WIDE-ANGLE DATA

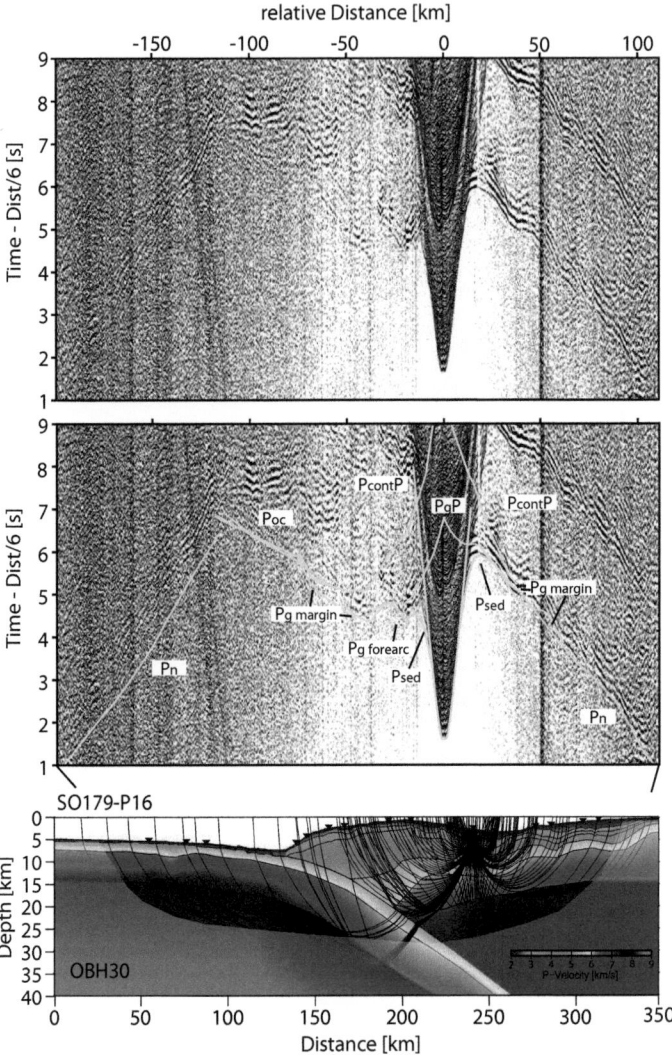

Figure 3.27: Forward model OBH30, profile SO179-P16.

3.3. RESULTS OF THE FORWARD AND THE INVERSE MODELING

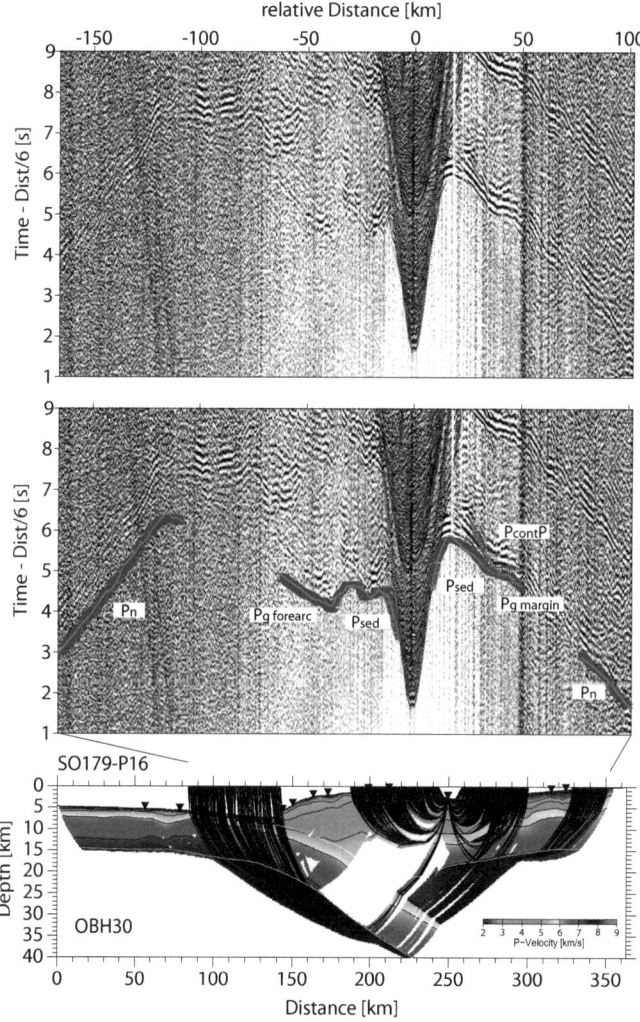

Figure 3.28: Inverted model of OBH30.

60 CHAPTER 3. MODELING THE SEISMIC P-WAVE VELOCITIES FROM WIDE-ANGLE DATA

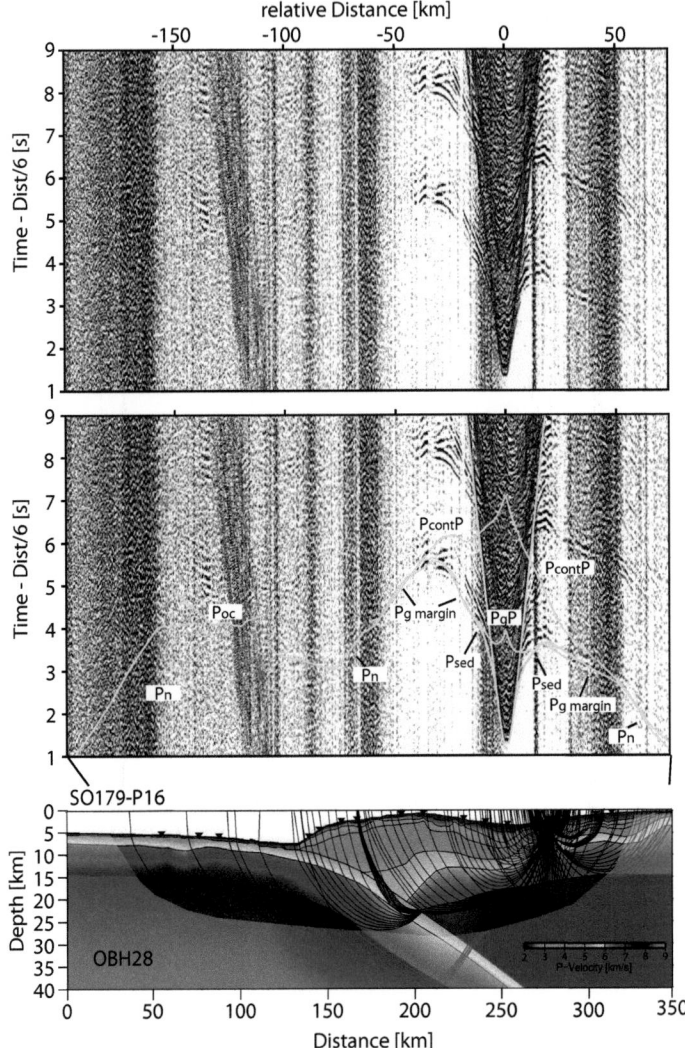

Figure 3.29: Forward model of OBH28, profile SO179-P16.

3.3. RESULTS OF THE FORWARD AND THE INVERSE MODELING

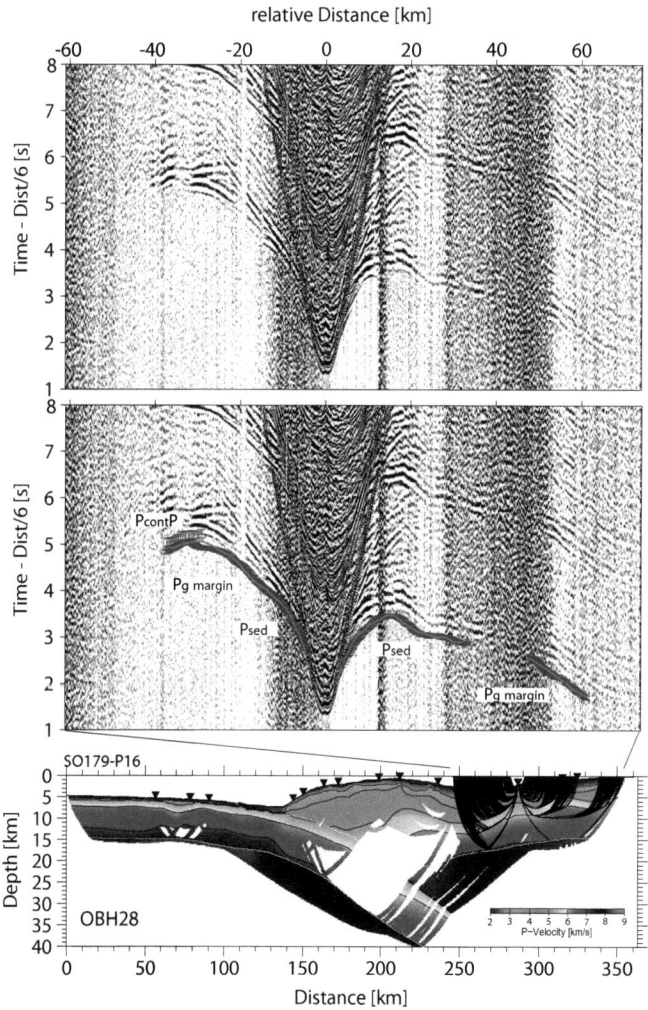

Figure 3.30: Inverted model of OBH28.

The lower boundary beneath the shelf and forearc basin becomes shallower on profile SO179-16, starting at a maximum depth of 15 km beneath the crest of the forearc high in the landward direction and then decrease to a minimum depth of 7 km below sealevel at the shelf. The velocity gradient of 6.0 - 7.1 km/s over the top of the subducting plate decreases in the landward direction to 6.7 - 7.2 km/s beneath the shelf. A second order layer boundary is displayed below this boundary with a velocity gradient of 7.2 - 7.9 km/s.

3.3.6 Shelf area

Data from the shelf area presented here in this section provide a very good wide-angle data set of OBH62 from the western profile SO179-P18. On the other hand the strikeline of profile SO179-P19 is located approximately 200 km landward of the deformation front on the shelf area of central Java and shall constrain the structural geometry in three dimensions between the two dip lines.

The station OBH62 (Fig's. 3.31 and 3.32) is located in a very shallow water depth of 632 m and displays a wide-angle data set which can be traced over the entire length of 350 km. The energy of the Bolt guns could be directly transferred to the subsurface and generated deeply penetrating rays. One common feature on the western profile SO179-P18 is the inherited structure which penetrates the shallow parts from the depth on profile-km 290. With the PcontP phase a crustal mantle boundary in a depth of 15 km could be verified (Fig's. 3.31 and 3.32). Rays leaving the mantle wedge (Fig. 3.31) are focused in the pinch out at profile-km 180, which is an undesired effect. However it could not be avoided, because in this very small model portion, the velocity gradients are unrealistically high. The oceanic crust is provided by the wide-angle data set between profile-km 100 and 150 by a low-velocity layer with a certain decrease in the velocities. The seismic amplitudes are decreased as well. The velocity at the deeper crust is 7.0 km/s, the mantle velocity is slightly 0.1 km/s higher (Fig's. 3.31 and 3.32). The differences from the forward model to the tomographic inversion is the highest below the forearc basin and the shelf area. The velocities in the tomography are 0.2 - 0.3 km/s higher compared to the forward model. This determines higher velocity uncertainties in this area.

This crust mantle boundary on profile SO179-P16 is positioned in the same depth of 15 km (Fig's. 3.31 and 3.32). The velocity contrast between the crust and mantle is higher compared to the eastern profile. The mantle velocities are 7.3 - 7.6 km/s.

Profile SO179-P19 should constrain the structural geometry. The seismic reflection data were also included in the forward modeling (Fig. 3.33). The layers are aligned subparallel to the seafloor and mainly horizontal. The superimposed inverted P-wave velocity field follows the seafloor as well. The seafloor exhibit some topographic undulations (Fig. 3.33), with troughs located at station OBH66 on profile-km 105 and at OBH74 at the end of the profile at profile-km 190. These are formed by river currents and are the origin for the sediment supply. The inherited structure at profile-km 50 could be resolved with higher velocities compared to the surrounding area and could represent a basement

3.3. RESULTS OF THE FORWARD AND THE INVERSE MODELING

high. The seismic amplitudes related to the basement high are dramatically decreased and the first arrivals could not be traced. This determines a gap in the first break picks related to this feature in the tomography (Fig's. 3.39 and 3.38). Due to the lack of stations located at this portion of the profile the model between profile-km 0 - 70 km SO179-P19 is poorly resolved. However, stations OBH65 to OBH68 exhibit data to model the layer thickness and velocities of the inherited structure.

Faults related to slumping features are displayed between profile-km 60 - 80 km. Between profile-km 100 to 160 the seismic section shows some blanking zones penetrating the basement. These are indications for fluid pathways, pleading for water saturated, approximately 3 km thick sediments.

Among others the stations OBH66 and OBH73 (Fig's. 3.36 and 3.38) show a good data quality yielding phases over the entire profile. The model beneath 10 km depth is poorly resolved. The forward model displays a slightly WE dipping layer, which could not be confirmed in the inversion.

64 CHAPTER 3. MODELING THE SEISMIC P-WAVE VELOCITIES FROM WIDE-ANGLE DATA

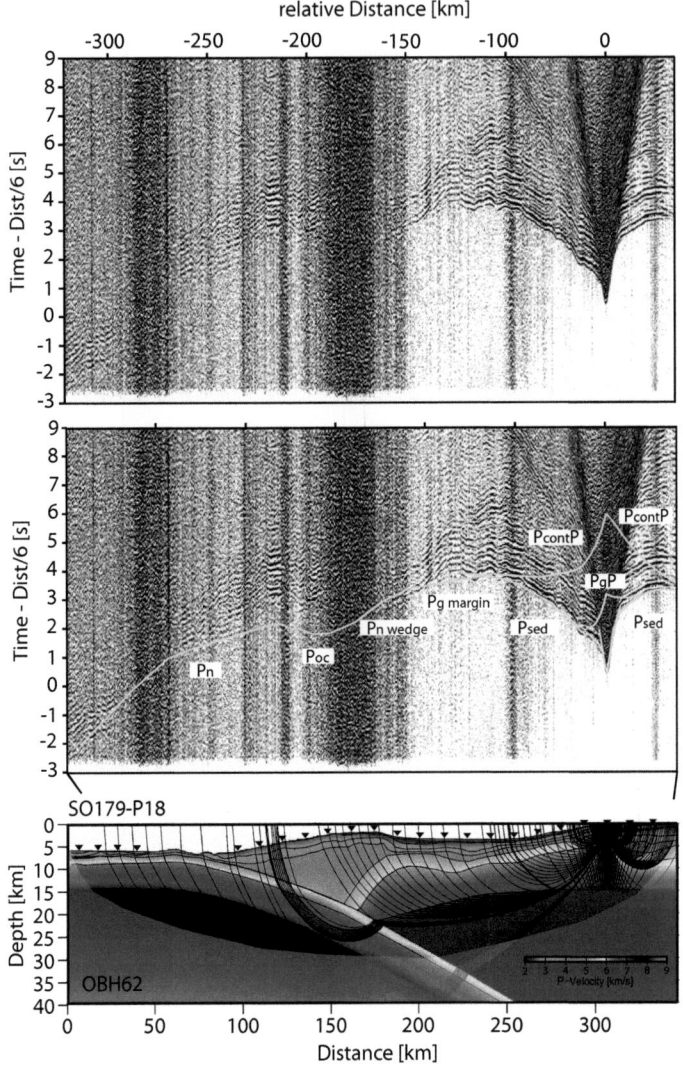

Figure 3.31: Forward model of OBH62, profile SO179–P18.

3.3. RESULTS OF THE FORWARD AND THE INVERSE MODELING 65

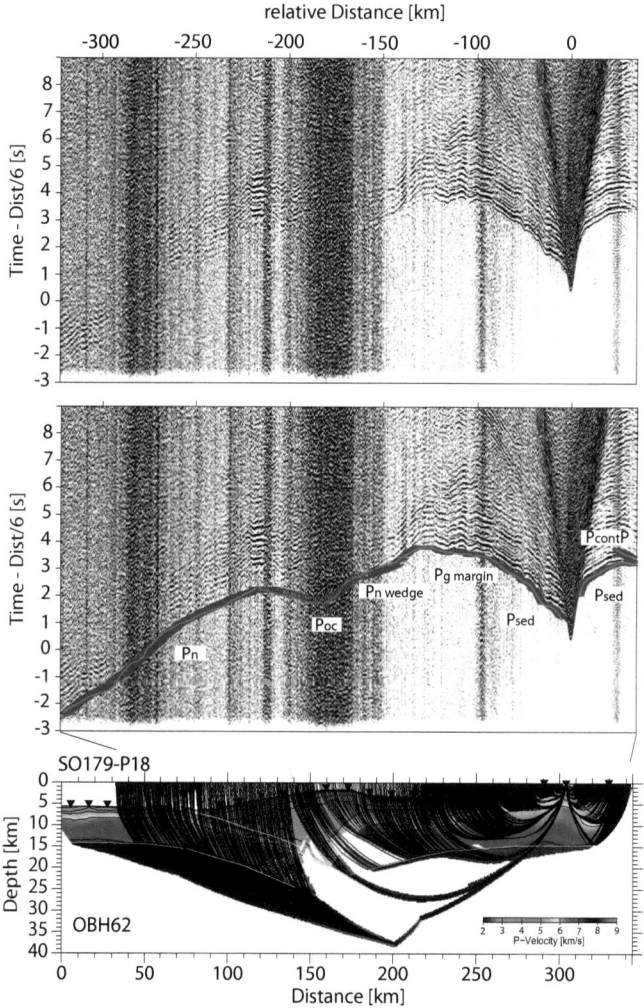

Figure 3.32: Inverted model of OBH62

66 CHAPTER 3. MODELING THE SEISMIC P-WAVE VELOCITIES FROM WIDE-ANGLE DATA

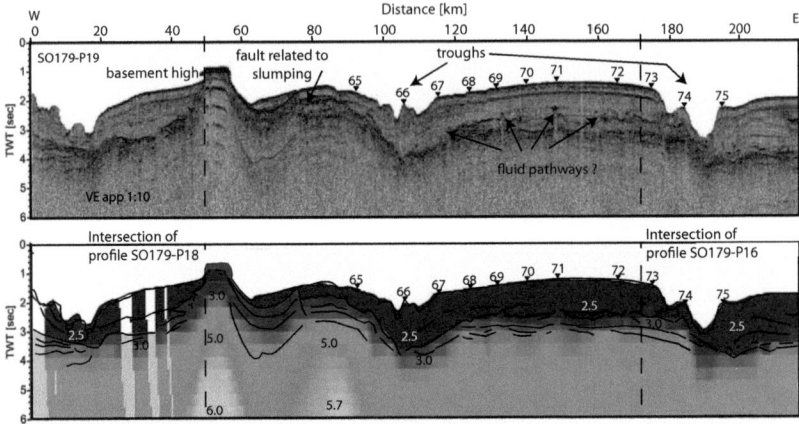

Figure 3.33: *Water migrated seismic record section of profile SO179-P16 (top) and interpretive line drawing (bottom). Station locations are indicated by triangles. The intersection of dip profiles SO179-P18 and SO179-P16 are marked by dashed lines. The tomographic velocity-depth model is superimposed and the seismic P-wave velocities [km/s] are notified as numbers.*

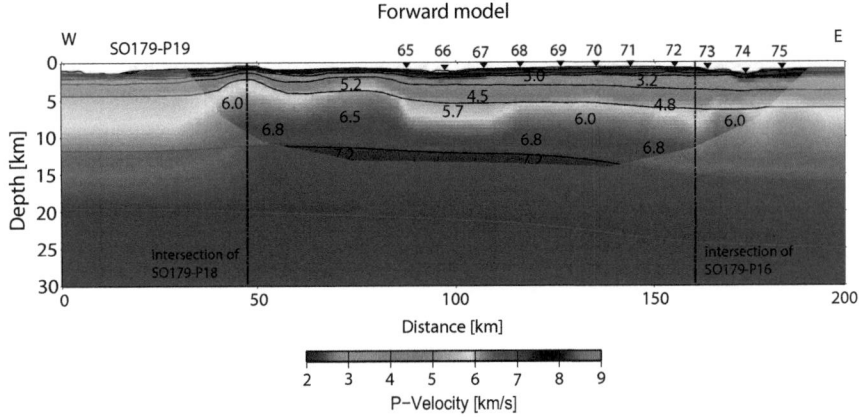

Figure 3.34: *Forward models of profiles SO179-P18 (top) and SO179-P16 (bottom). The P-wave velocities are color coded and additionally marked by numbers [km/s]. Station locations are indicated by triangles.*

3.3. RESULTS OF THE FORWARD AND THE INVERSE MODELING

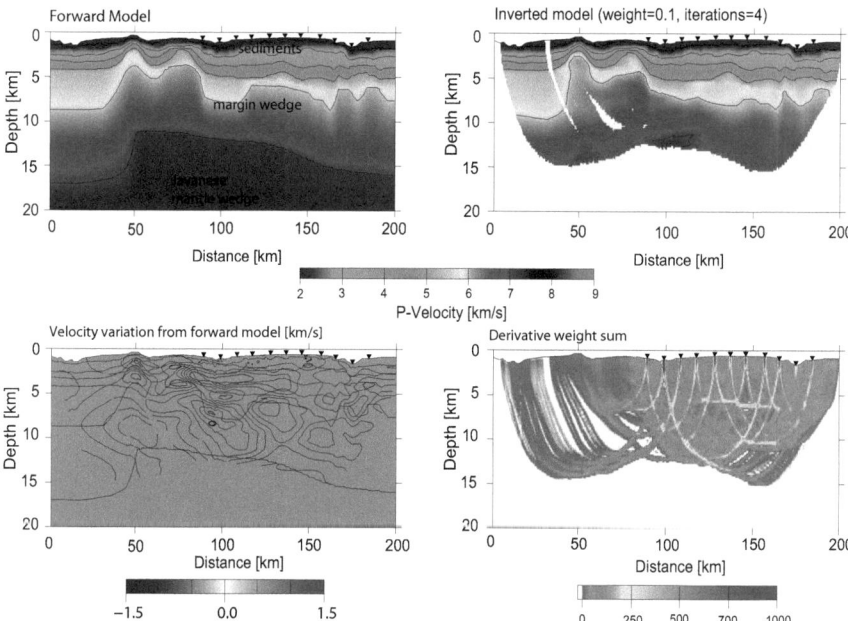

Figure 3.35: *Velocity model updates for the final model of profile SO179-P19 are less than < 1.9% after the first iteration. Therefore a velocity damping was not applied. After five iterations model updates became < 0.1%. The RMS misfit of all refracted rays after last iteration is 56 ms. The final χ^2 value is 1.23*

68 CHAPTER 3. MODELING THE SEISMIC P-WAVE VELOCITIES FROM WIDE-ANGLE DATA

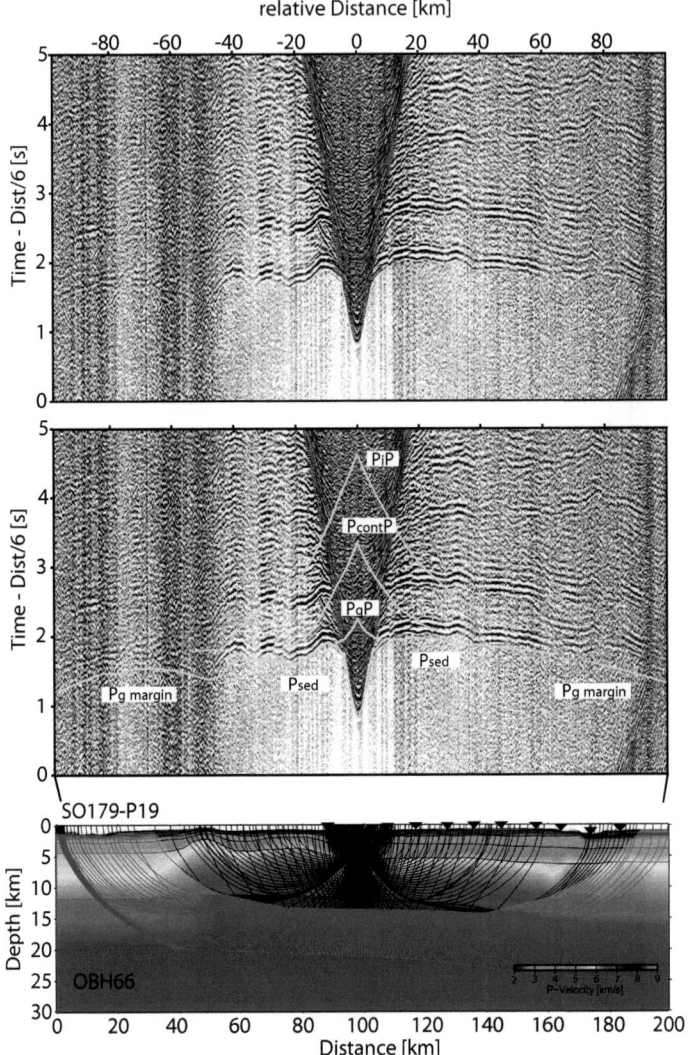

Figure 3.36: Forward model OBH66, profile SO179-P19.

3.3. RESULTS OF THE FORWARD AND THE INVERSE MODELING

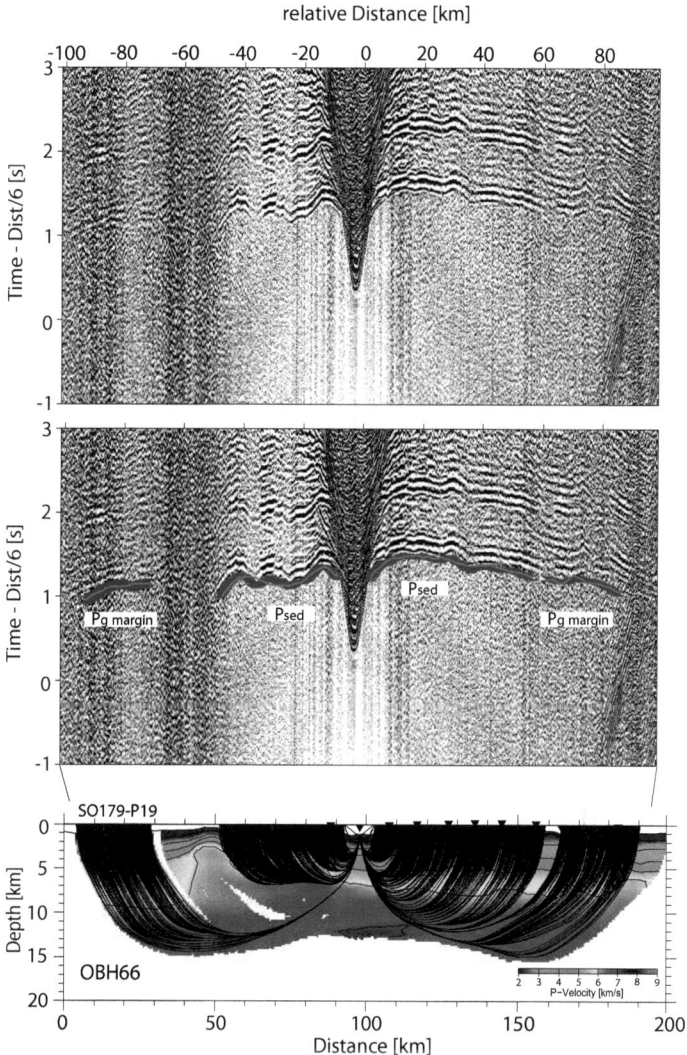

Figure 3.37: Inverted model of OBH66

70 CHAPTER 3. MODELING THE SEISMIC P-WAVE VELOCITIES FROM WIDE-ANGLE DATA

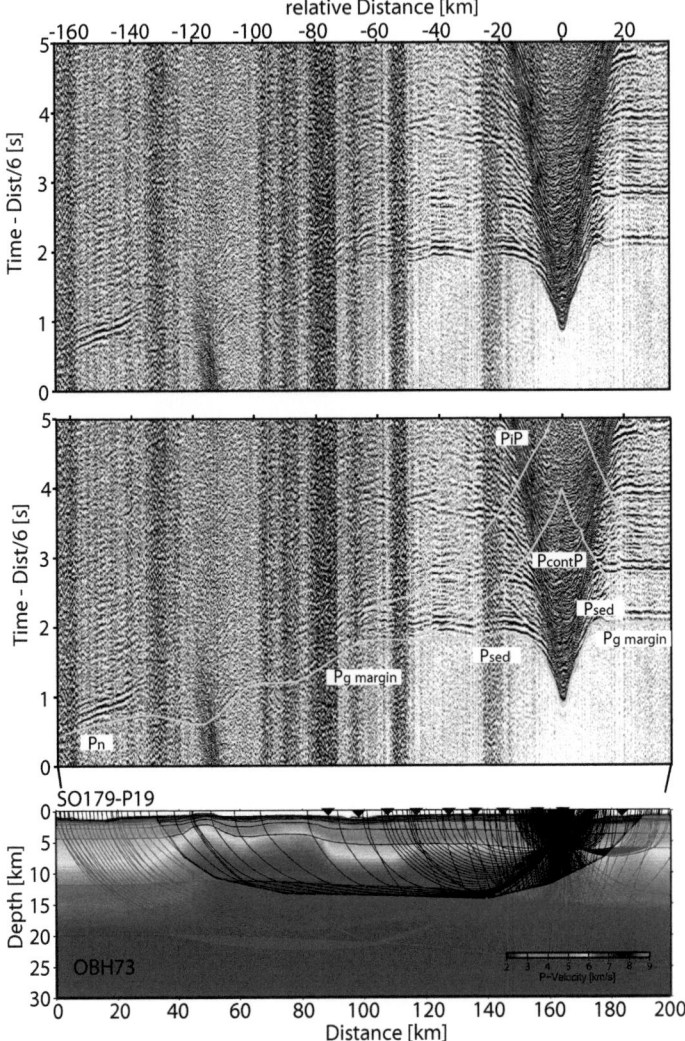

Figure 3.38: Forward model OBH73, profile SO179-P19.

3.3. RESULTS OF THE FORWARD AND THE INVERSE MODELING

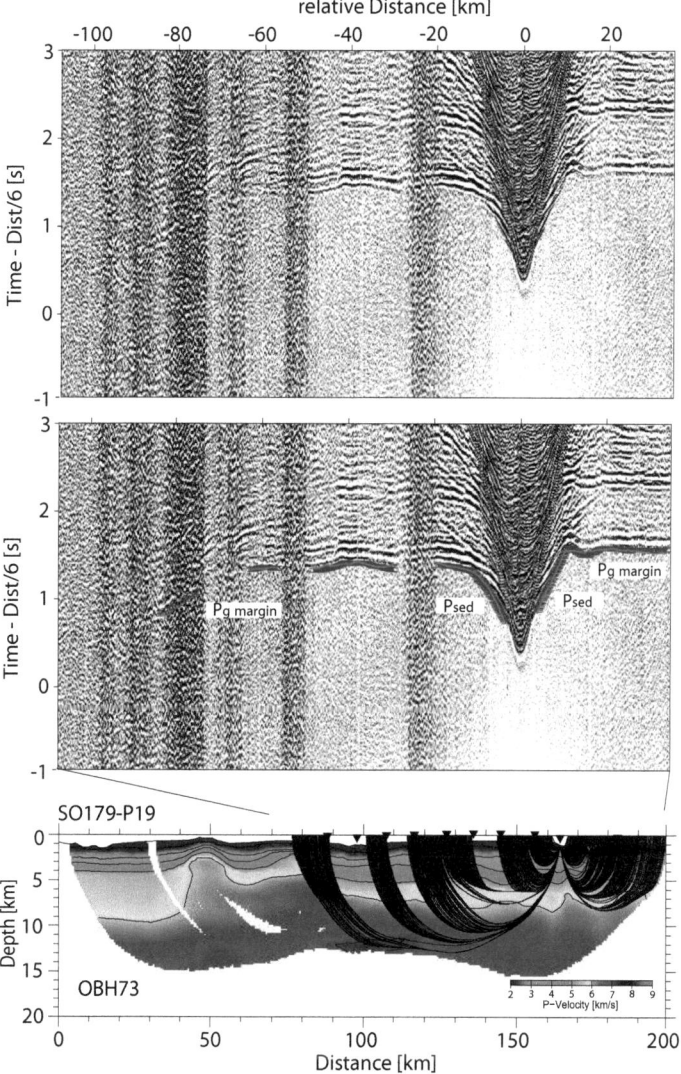

Figure 3.39: Inverted model of OBH73

3.3.7 Model uncertainties and sensitivity tests

The uncertainties in the velocity-depth modeling are linked to the traveltime data misfits (which correspond to the data vector of Eqn. 3.5) of the tomographic inversion and the uncertainties of the velocity-depth model (according to the model vector of Eqn. 3.5). The discussion in this chapter is mainly based on the inversion results. The forward model does not provide any uncertainty values. All traveltime picks have a residual $\Delta t = t_{obs} - t_{calc}$, which is the difference between observed and calculated traveltimes after the inversion. Figure 3.40 displays the traveltime residuals for the first (left) and the last (right) iteration of the tomographic inversion. The residuals of the first iteration corresponding to the uncertainties of the forward model. This figure displays the good results of the forward model which could be refined and improved by the tomography: the shape of the normal distributed function has a smaller interval around the mean value (this determines a smaller standard deviation) and less data outliers. Data fitting is strongly linked with data uncertainties (Zelt, 1999). The total traveltime misfit consists of the pick uncertainties and the traveltime residuals (Tab. 3.2). The near offset phases have in general a small picking uncertainty due to the strong seismic amplitudes with a few gaps. The far offset phases have a higher pick uncertainty, the phases could not be traced completely. The reflected phases are superimposed by the refracted phases and the onsets could not be clearly identified (Tab. 3.2)

phase	pick uncertainties [ms]
P near offset (≤ 70 km)	50
P far offset (≥ 70 km)	80-100
Pn	80-100
PmP	80-100
PtocP	80-100
PcontP	80-100

Table 3.2: Picking uncertainties

With the residuals $\Delta t = t_{obs} - t_{calc}$ follows:

$$t_{mean} = \frac{1}{n}\sum_{i=1}^{n} \Delta t_i = \frac{\Delta t_1 + \Delta t_2 + .. + \Delta t_n}{n} \quad (3.13)$$

$$t_{STD} = \sqrt{\frac{1}{n-1}\sum_{i=1}^{n}(\Delta t_i - t_{mean})^2} \quad (3.14)$$

$$t_{RMS} = \sqrt{\frac{1}{n}\sum_{i=1}^{n}\Delta t_i^2} = \sqrt{\frac{\Delta t_1^2 + \Delta t_2^2 + .. + \Delta t_n^2}{n}} \quad (3.15)$$

t_{mean} represents the arithmetic mean value of all the residuals of a tomographic inversion, including the standard deviation t_{std} of a normal distributed data set. The root mean square (RMS)

3.3. RESULTS OF THE FORWARD AND THE INVERSE MODELING

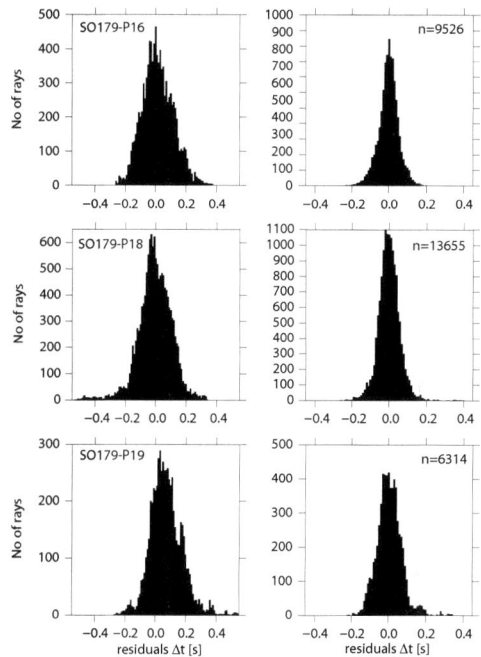

Figure 3.40: Distribution of traveltime residuals $\Delta t = t_{obs} - t_{calc}$ values after first (left) and last iteration (right). First iteration corresponds to the forward model. Only refracted phases to determine the velocity model

value t_{RMS} describes the quadratic mean value and accounts for a stronger influence of higher residuals. The tomographic inversion provides results with fairly well RMS values of 53 - 63 ms (Tab. 3.3) of refracted phases, which is reasonable well within the pick uncertainties (Tab. 3.2). The inversion of reflected phases provides higher RMS uncertainties of 28 - 101 ms, due to the much lower number of picks and the higher picking uncertainties. However, only a very small number of outliers with values $\geq 0.2s$ is increasing the RMS value (Fig. 3.40). The arithmetic mean values are 3.2 - 56.3 ms and the maximum standard deviation is $\pm$ 9.9 ms (Tab. 3.3).

For judging the quantitative resolution and accuracy of the model, resolution tests are applied. The inversion is tested on the real data set with a checkerboard test and on synthetic datasets. All tests are applied to reconstruct a pattern of known anomalies using the same source-receiver geometry and identical traveltime data, as used in the inversion. A solution is reliable, when a

74 CHAPTER 3. MODELING THE SEISMIC P-WAVE VELOCITIES FROM WIDE-ANGLE DATA

tomography traveltime uncertainties	SO179-P16	SO179-P18	SO179-P19
t_{mean} P phases [ms]	14.3 ± 5.3	4.3 ± 4.9	56.3 ± 8.5
t_{RMS} P phases [ms]	52.88	55.86	63.60
min/max Δt [s]	-0.23/0.21	-0.25/0.40	-0.22/0.34
no of reflected phases	9526	13655	6314
t_{mean} PmP [ms]	4.2 ± 2.7	3.2 ± 1.8	-
t_{RMS} PmP [ms]	52.9	28.0	-
min/max Δt [s]	-0.13/0.30	-0.07/0.09	-
no of PmP's	476	237	-
t_{mean} PtocP [ms]	29.0 ± 5.7	43.0 ± 5.7	-
t_{RMS} PtocP [ms]	99.1	80.0	-
min/max Δt [s]	-0.27/0.25	-0.17/0.41	-
no of PtocP's	346	593	-
t_{mean} PcontP [ms]	17.8 ± 9.9	37.3 ± 3.8	-
t_{RMS} PcontP [ms]	101.1	72.0	-
min/max Δt [s]	-0.23/0.23	-0.25/0.11	-
no of PcontP's	107	577	-

Table 3.3: Uncertainties of the final P-wave velocity models of the three profiles.

known model structure with similar length scale to the final solution can be recovered using similar ray paths.

An alternating pattern of low and high velocity anomalies is superimposed to the derived models in a checkerboard test. The length scales of the anomalies should be similar to the smallest wavelength structures in the model. For this approach a checkerboard pattern with a cell size of 45x8 km and a velocity anomaly of ±0.55 km/s is applied to the velocity models. Figure 3.41 shows the resulting pattern after four iterations. The area within the station coverage in the uppermost parts of the model, the shape pattern is recovered fairly well. Due to a station gap at the trench the shapes of the checkers are blurred on profile SO179-P16, whereas the shape recovery below the trench is quite good on SO179-P18. On both profiles the recovery is reasonable to a depth of 10 km bsf.

The retrieved amplitudes reach more than 90% at the uppermost parts of the models. The checkerboard test fails at the lowermost parts and the edges of the model due to the lack of raycoverage. The resulting checkerboard pattern of the northern profile SO179-P19 is resolved very well between profile-km 100 to 200 into a depth of 7 - 10 km. The westernmost portion at the intersection to the western profile is poorly resolved, caused by missing stations. However an alternating pattern of positive and negative checkers can be detected in the uppermost part of the model.

The calculated DWS vector provides qualitative information about the ray coverage (Fig. 3.42). High DWS values corresponds to a denser sampling of rays and an accumulation of higher quality ray paths with smaller pick uncertainties. The well resolved regions in the checkerboard tests coincide

3.3. RESULTS OF THE FORWARD AND THE INVERSE MODELING

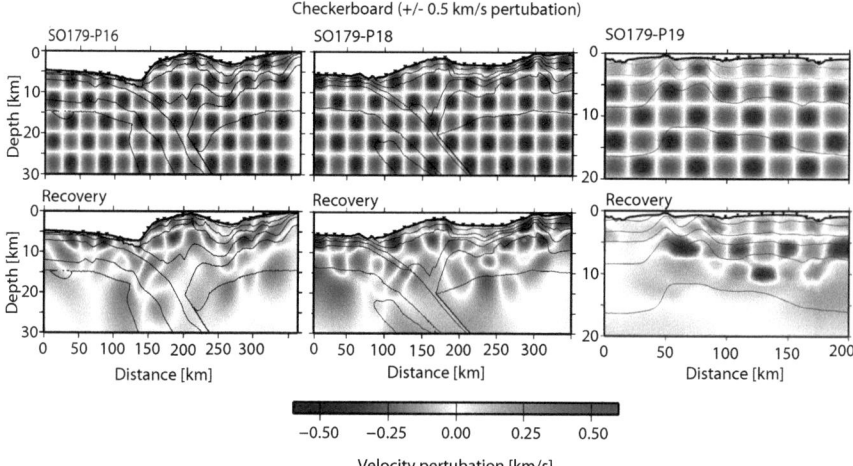

Figure 3.41: Checkerboard test for all three profiles. An alternating pattern of 45x8 km rectangles with values of ±0.55 km/s positive and negative velocity perturbations was applied to the final models (top). The recovered anomaly pattern after four iterations is displayed below.

with high DWS values and a well resolved model with a minimum velocity uncertainty.

In a further test, a synthetic data set is calculated with the same source and receiver distribution for a perturbed velocity model (Fig. 3.43). Six alternating positive and negative synthetic anomalies are placed in a depth of 10 km into the final model (Fig. 3.43).

Synthetic anomalies are positioned at key locations of the velocity-depth models SO179-P16, SO179-P18 and SO179-P19. Regions of interest are the forearc high, the forearc basin, the oceanic mantle and the lowermost portion of the upper plate. (Fig. 3.43). The forearc high close to the trench is of special interest, to determine the scale of the frontal prism and the location of the backstop.
For this purpose Gaussian anomalies of $Aexp(-(x-x_0)^2/L_h - (z-z_0)^2/L_v)$, with amplitude A, location x_0, z_0, horizontal length of L_h and a vertical length of L_v respectively, were placed into the final model. The amplitudes reaches ±0.4 km/s. The resolved synthetic shallower anomaly amplitudes could be recovered to almost 100% (Fig. 3.43). The P-wave velocities at shallow model structures could be determined with a maximum error of ± 0.1 km/s on both profiles. The synthetic anomaly in the lower part of the upper plate could be recovered in shape and 75% amplitudes on profile SO179-P16. The anomaly on profile SO179-P18 could only be recovered with 50%, which results in a maximum error of ±0.2 km/s. The mantle velocity anomalies on profile SO179-P18 could also be recovered with 50% of the maximum amplitude, which results in a maximum error of ±0.2 km/s for the mantle velocities.

76 CHAPTER 3. MODELING THE SEISMIC P-WAVE VELOCITIES FROM WIDE-ANGLE DATA

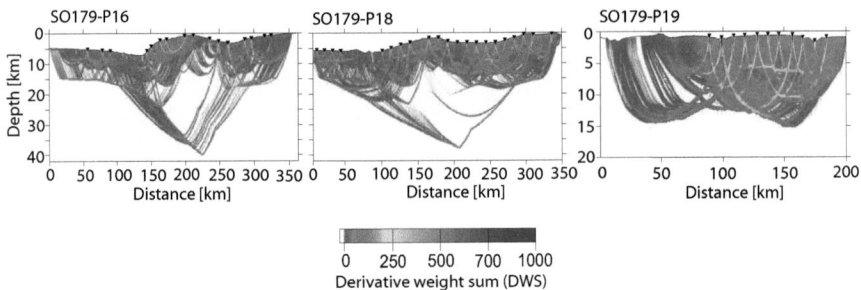

Figure 3.42: DWS plot gives an indication of ray coverage and high quality ray paths. Red and yellow ray paths have a high dws value.

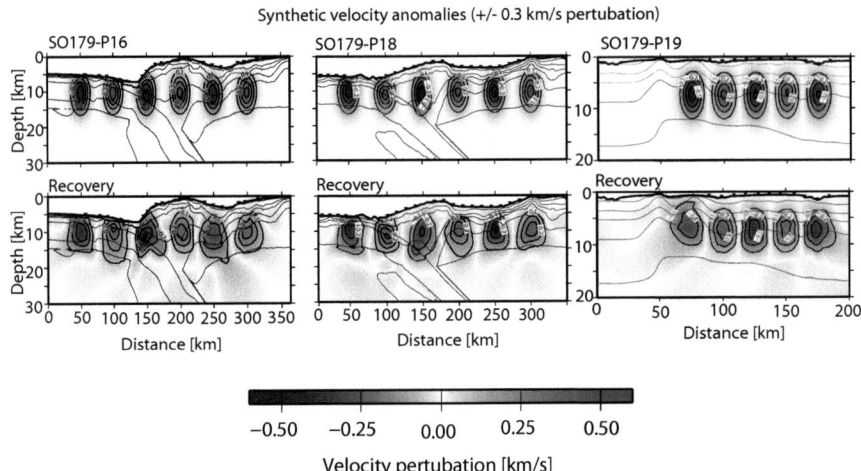

Figure 3.43: Alternating positive and negative synthetic anomalies with a maximum amplitude of ± 0.4 km/s. The thin iso-velocity lines in the background are displayed for a better orientation to the model structures.

3.3. RESULTS OF THE FORWARD AND THE INVERSE MODELING

The success in retrieving the synthetic anomaly patterns suggests that a similar structure can be resolved in the experiment. The tomographic solution has a minimum sensitivity of ± 0.2 km/s for small velocity perturbations within the area of station coverage. At a depth below 15 km the resolution of large-scale structures suffers from the decreasing raycoverage with increasing depth.

3.4 Conclusion

Offshore central Java a thickened oceanic crust of 9 km (SO179-P18) - 10 km (SO179-P16) km, covered with a thin layer of sediments with a thickness of 500 m is subducting with a dip-angle of 10° below the Eurasian plate. The oceanic mantle velocities of 7.9 - 8.0 km/s are 'normal' on profile SO179-P16, and decreased P-wave mantle velocities are recognized on the western profile (7.7 - 7.9 km/s). On the eastern profile SO179-P16, the velocity contrast between the outer and inner wedge, which is defined by the backstop boundary is smeared and not as clearly defined as on the western profile. It seems to be influenced by a formerly subducted seamount, which determines the uplift of the entire forearc high. An inherited basement high with increased seismic velocities on profile SO179-P18 is located to the adjacent undeformed forearc basin with flat lying sediments. The layering across profile SO179-P19 is mainly horizontal with slightly E-W dipping interfaces. The P-wave velocity model fits the merged intersection of the dip-lines (Fig. 3.44).

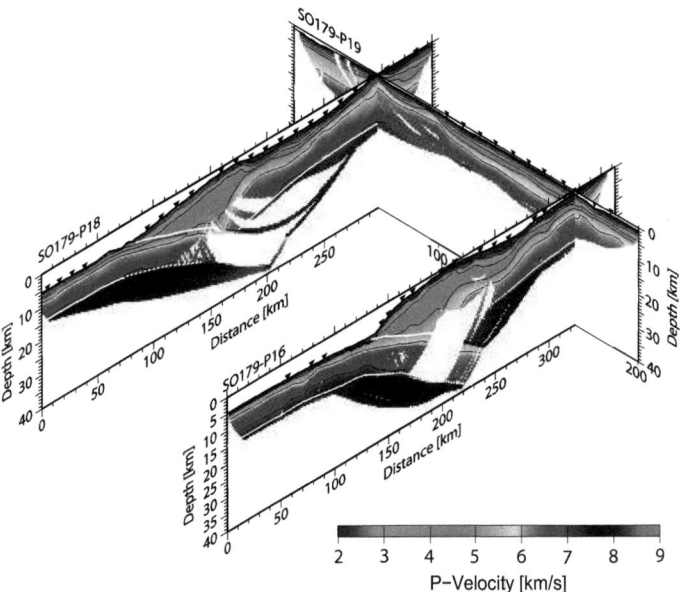

Figure 3.44: Merged tomographic models in a perspective 3D view. The velocity slices are based on the joint tomographic inversion of reflected and refracted phases. The strike line SO179-P19 includes only refracted phases for the inversion. The velocity fields at the intersections of the independently inverted dip lines to the strike line fit very well.

Chapter 4

Gravity

A two-dimensional gravity analysis provides further constraints on the structural models. The gravity models were developed with MacRay (Luetgert, 1992) from a priori information of the structures obtained by the seismic velocity-depth models (refer to previous Chapter). Due to the non-uniqueness of gravity modeling it is important to utilize the velocity-depth model. Therefore the geometry of the layer boundaries was directly used for the gravity modeling. Velocities can be converted using velocity-density relations (Ludwig et al., 1970), which is dependent on the pore space in rocks in the different continental and oceanic environments.

The calculated gravity profiles are compared to ship data acquired by the Federal Institute of Geosciences and Natural Resources (BGR), during SO179 cruise (Kopp and Flueh, 2004) (Fig. 4.1).

The dominating anomalies of the gravity map (Fig. 4.1) are mainly influenced by topographic features south off Java. The oceanic crust in the south shows positive gravity values from 0 - 40 mGal. A seamount anomaly located at 12°S/ 109°E provides a local gravity maximum of 70 mGal. The southern edge is dominated by positive anomalies caused by the outer rise of the downgoing lithosphere. The trench is indicated by a gravity low of -150 mGal, which occurs over a deeper trench (7500 m) compared to Sumatra with a gravity low of -60 mGal (corresponding to 6000 m). The forearc high between 9° S and 10° S with its shallow water depth shows local positive anomalies in a generally negative environment. The forearc basin with its low density sediments is located north of the forearc high with an elongated minimum parallel to the trench. A strong positive anomaly of up to 200 mGal dominates the southern shoreline off central Java, which can not only be explained with the topography but also with a higher density and a shallow mantle wedge in the subsurface.

The gravity models are kept as simple as possible with constant densities in the corresponding units and only a few gravity gradients in the oceanic mantle of the western profile and in the outer wedge. Gravity anomalies vary up to 300 mGal over the entire dip lines (Figs.' 4.3 and 4.4). The calculated gravity fit the observed data within a range of less than 10 mGal. To take into account side effects, the model space is much bigger compared to the velocity-depth models. The model space was extended ± 100 km along dip (SO179-P16/P18) and strike direction (SO179-P19) and 20 km in depth to 50

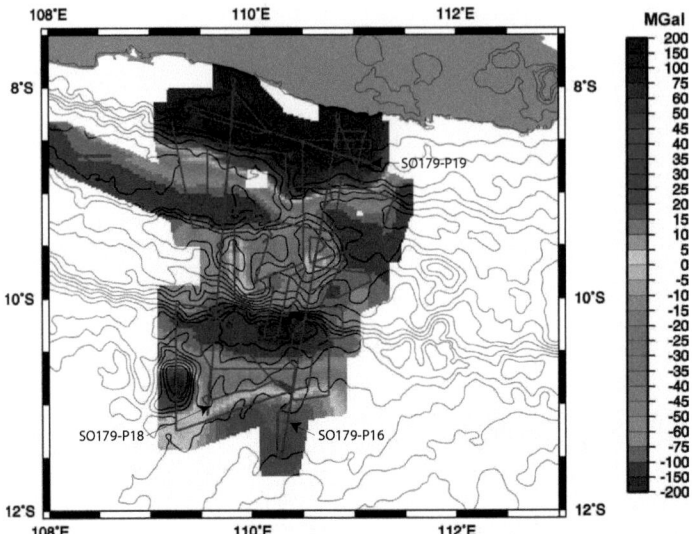

Figure 4.1: Free-air gravity anomalies in the survey area. Ship tracks are in green. This map was compiled by the Federal Institute of Geosciences and Natural Resources (BGR).

km in total.

Based on Ludwig et al. (1970) an average density of 2.2 g/cm^3 for the unconsolidated sediments with velocities < 3 km/s was assigned to the upper sedimentary layers. Establishing the velocity-depth modeling results, the western profile SO179-P18 provides a much thicker low density layer on the oceanic crust, influenced by a volcanic chain and the neighboring seamount at 11°S/ 109°E. The gravity on the outer rise of the western profile SO179-P18 is about 50 mGal decreased comparing to the eastern profile, which could be a result of the flexure of the oceanic lithosphere caused by the load of the seamount at 11°S/ 109°E (Fig. 4.2).

The large size of the seamount and its corresponding heavy mass causes a moat structure around the seamount, where sediments are trapped and accumulated. The mass of the seamount would also generate intense faulting, which reduces the densities (and the corresponding P-wave velocities) in the entire oceanic crust, including the altered mantle (Contreras-Reyes and Osses, 2010).

The accreted sediment load at the forearc high close to the trench is larger compared to the eastern profile. Accreted sediments with lower densities prevail on SO179-P18, which classifies this profile more as an accretionary subduction system compared to profile SO179-P16. The sediment load on profile SO179-P16 is not significant on the outer rise until the rising slope of the forearc high. The uncompacted sediment layer increases in thickness in landward direction. Both dip lines display an

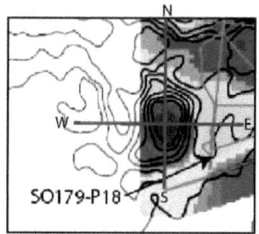

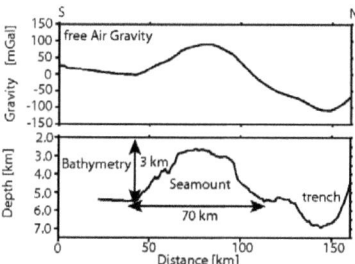 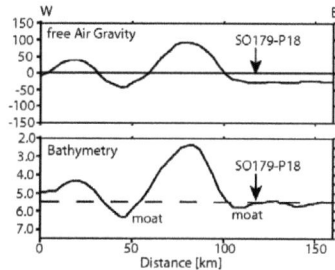

Figure 4.2: Top: Blow up of of Fig. 4.1 with the region of the located large seamount around 12°S and 109°E. Red lines display the NS- and WE trending traces of bathymetry and gravity data below. Bottom, left: NS trending trace of the free air gravity corresponds to the shape of the high resolution bathymetry of the seamount with its large dimensions. Right: WE trending trace with free air gravity and the bathymetry (lower resolution from global ETOPO dataset), with its moat around the seamount.

uppermost sediment layer with varying thickness, corresponding to first order discontinuities and the P-wave velocities of the forward model. The sediments reach a maximum thickness of approximately 3.5 km at the forearc basin.

For the oceanic crust a uniform density of 2.90 g/cm^3 as suggested by Carlson and Raskin (1984) was applied for profile SO179-P16, whereas the western profile requires lower densities of 2.85 g/cm^3. This is consistent with lower P-wave velocities in the velocity-depth model. The density of the oceanic mantle of profile SO179-P18 has a significantly lower density of 3.32 g/cm^3 compared to the eastern profile, with mantle densities of 3.37 g/cm^3 yielding a good fit beneath igneous oceanic crust on profile SO179-P16. However, trying to adjust and equalize the differences in the mantle densities on both dip lines results in a gravity misfit of up to 50 mGal and does not match the observed gravity (Fig. 4.3). These density values are supported off western Java by the GINCO profile (Kopp et al., 2002). The P-wave velocities of SO179-P16 have 'normal' values of 8.0 km/s, whereas the western profile has reduced velocities of 7.8 km/s. Therefore a density gradient of 3.32 g/cm^3 at the top and 3.35 g/cm^3 was applied (Barton, 1986) and implies an altered oceanic mantle on the western profile.

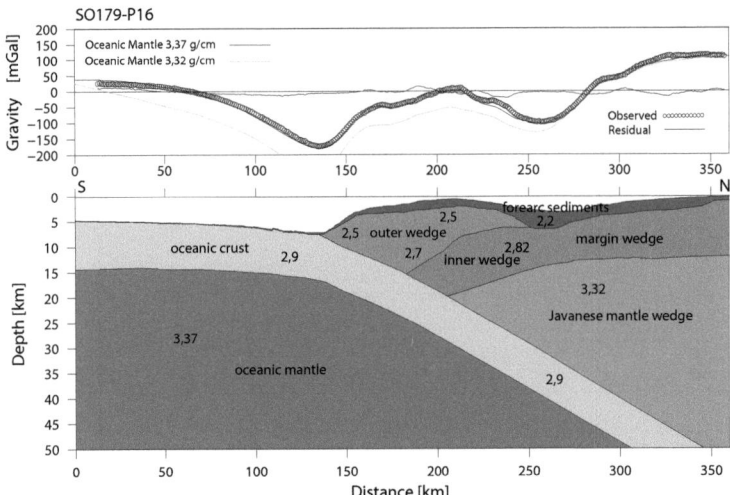

Figure 4.3: Calculated free air gravity of profile SO179-P16 (top). Applying a lower mantle density of 3.32 g/cm^3 (grey line) like on the western dip-line, does not match the observed gravity (circles). The calculated gravity fits the observed within a range of ± 10 mGal (residual gravity dashed line). Tectonic units (bottom) are color coded with constant density values.

For the deeper forearc structures a higher degree of compaction and metamorphism require densities ranging from 2.5 - 2.7 g/cm^3. The same P-wave velocities in the inner wedge on both dipping profiles require the same densities. The gravity in the forearc section is influenced by the shallow water depth and the older accreted and metamorphosed sediments. The inner wedge and the margin wedge requires a density of 2.85 g/cm^3 to fit the observed gravity. The gravity response on the landward part of the forearc basin is mainly influenced by the shallow mantle which reaches a depth of about 15 km. The Javanese mantle wedge fits the observed gravity with a constant value of 3.32 g/cm^3. The shallow mantle is in agreement with the velocity-depth model and with the results presented by Grevemeyer and Tiwari (2006). The Javanese mantle wedge densities could be verified by Shulgin et al. (2010) on the neighboring SINDBAD profile. Whereas the gravity model of the GINCO profile off western Java requires higher Javanese mantle wedge densities of 3.37 g/cm^3 (Kopp et al., 2002).

The gravity profile of SO179-P19 (Fig. 4.5) is divided into a gently west to east dipping layering of sediments, crustal units and the Javanese mantle. At profile-km 50 (cross-line to profile SO179-P18) the crust penetrates the shelf sediments. This profile matches the density values of the dip-lines and thus support the velocity-depth models of chapter 3.

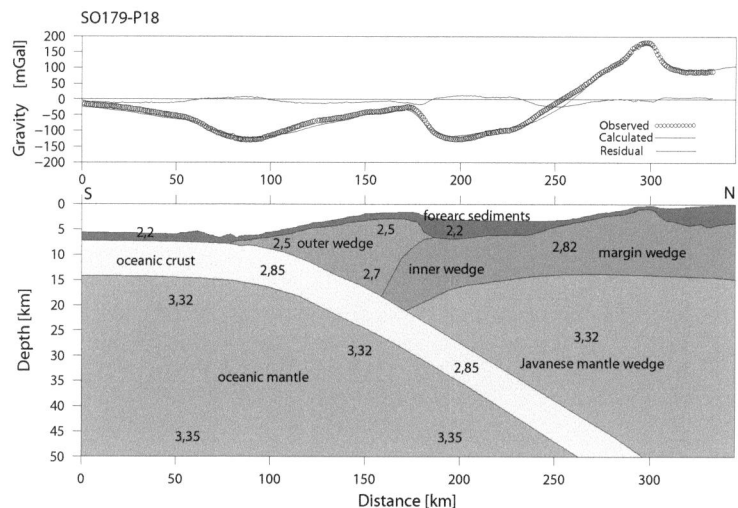

Figure 4.4: Calculated free air gravity of profile SO179-P18.

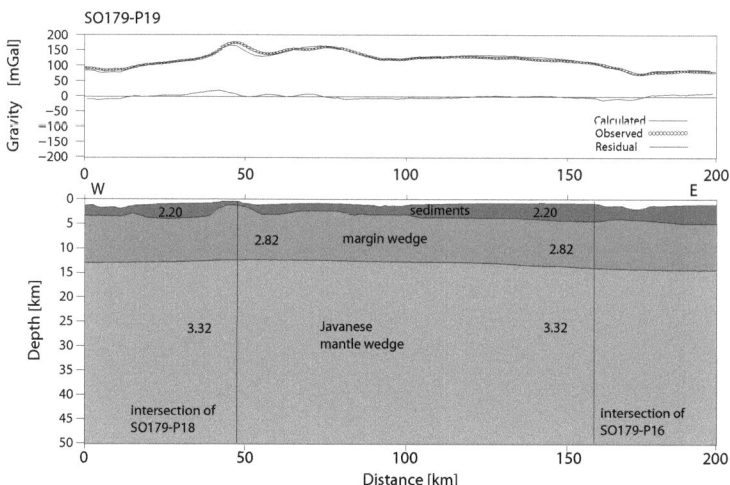

Figure 4.5: Calculated free air gravity of profile SO179-P19.

Chapter 5

Discussion

A discussion of the results from previous sections based on the combined forward- and inverse modeling to determine a detailed velocity-depth model of the three wide-angle profiles offshore central Java is presented here. The interpretation will be completed with the results of the gravity modeling, the reflection seismic and bathymetric data.

This section is divided into the description of the incoming oceanic lithosphere and interplate processes. The results of the MERAMEX project will be compared with the GINCO and SINDBAD dataset to characterize the forearc region offshore the island off Java. The passive seismic dataset, containing approximately 500 earthquakes located with the temporary onshore seismometers by the German Research Center for Geosciences (GFZ), will be presented. The onshore tomography results from Wagner et al. (2007) and Koulakov et al. (2007) are completing this chapter.

5.1 Oceanic lithosphere and the Christmas Island seamount province

The incoming plate off central Java is covered by approximately 500 m of pelagic/hemipelagic sediment (500 ms TWT, Figure 3.11), which is 66% less than that on the western Java transect where average sedimentary thickness is about 1500 m (Kopp et al., 2002). The increasing distance from the Bengal fan and the curvature of the trench offshore Sunda Strait cause a decrease of the trench fill to less than 1 km thickness. The trench varies between ≥ 5600 and ≤ 7000 m in depth between the Sunda Strait at 105° E and eastern Java at 115° E. The trench is devoid of sediments with the exception of local ponded, flat-lying accumulations, which reach hundred meters in thickness (Fig. 3.11).

Figure 5.1 displays the structural interpretation of the investigated velocity-depth models in this study. The velocity-depth function of the oceanic crust is represented with two depth slices AA' and BB' in Figure 5.1, where the tomographic model is well resolved. Global compilations of

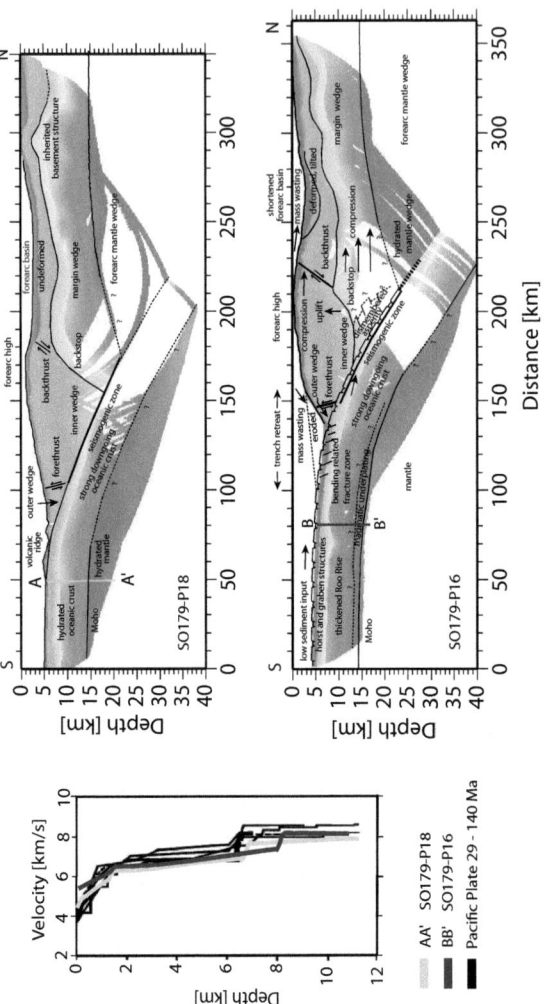

Figure 5.1: Interpretation of the modeling results. The western profile SO179-P18 is displayed at the top, the eastern profile SO179-P16 at the bottom. The solid lines indicate an interpretation of the structural boundaries. The color coded P-wave velocities are displayed in the background with a reduced opacity. The velocity-depth functions of these models (yellow, blue) are compared with a global compilation of v(z) of the Pacific plate (black) by White et al. (1992).

5.1. OCEANIC LITHOSPHERE AND THE CHRISTMAS ISLAND SEAMOUNT PROVINCE

seismic refraction velocity-depth functions for 29 - 140 Ma old Pacific oceanic crust by White et al. (1992) are compared with the velocity-depth function of the central Java profiles. This compilation is referred to 'normal' oceanic crust. The hydration is only limited to the permeable upper lava pile and the oceanic lithosphere is positioned away from anomalous regions (White et al., 1992).

The western profile velocity-depth distribution displays reduced velocities from 4 - 6 km/s in layer 2 compared to the eastern profile SO179-P16 at the first two depth-kilometers. Figure 4.2 shows a large positive gravity anomaly, which corresponds to a seamount with a base diameter of 70 km, a height of 3 km and an approximated conic volume of 3850 km^3. The upper layer of reduced velocities implies an extrusive origin, composed of volcanic debris in the vicinity of the volcanic ridge. The flexural stress on the oceanic crust caused by the depression of the seamount leads to a activation of zones of crustal weakness (Ranero et al., 2003). This implies a fault pattern of inherited defect structures originating from the emplacement of the volcanic ridge, which are reactivated in the near trench setting. Bending related-, and flexural faulting increases the permeability in the oceanic crust and reduces the P-wave velocities significantly. A downflexing of the lithosphere could not be resolved due to the position on the outer rim of the moat structure (Fig. 4.2). Seamounts are compensated differently at depth, which is dependent on whether they have been formed on a strong plate in a plate interior or on a weak plate near a mid-ocean ridge (Koppers and Watts, 2010). This determines different states of crustal buoyancy as they enter the trench.

An important factor controlling the local hydrological system is the thickness of the sediments. Hydration is lower in well sedimented margins (Ranero et al., 2003). Basement outcrops allow hydrothermal fluids to bypass the sediments at the seafloor. This increases the water content in the oceanic crust and decrease the P-wave velocities, which determines an alteration of the oceanic lithosphere. Tectonic faulting may allow fluids trapped in pore spaces to enter lower oceanic crust and perhaps even the upper mantle (Grevemeyer and Tiwari, 2006). Since water will alter mantle peridotite to serpentine, pervasive fracturing of the entire oceanic crust is suggested. The oceanic mantle velocity of 8.0 km/s (Fig. 5.2) at the eastern profile is higher compared to mantle velocities of 7.7 km/s at the western profile. The lower P-wave mantle velocities can be the result of the alteration process of peridotite.

We follow the approximation formula of Carlson and Miller (2003) to estimate water content in partially serpentinised peridotites in the mantle: $w(\%) \approx -0.33\Delta V$. w is the water content (in weight percentage) and ΔV is the percent difference between observed velocity and the unaltered velocity in peridotite. The degree of serpentinisation and the water content of partially serpentinised peridotites can be related to the seismic P-wave velocity. In partially serpentinised ultramafic rocks, a decrease of 1 % of P-wave velocity corresponds to a increase in serpentine of 2.4% and a 0.3% increase in water content (Carlson and Miller, 2003). A 3.7% (7.7 km/s mantle velocities) decrease in P-wave mantle velocities (based on average upper mantle velocities of 8.0 km/s) results in 9% serpentine and roughly 1% water content in the upper oceanic mantle. These values are comparable to the oceanic

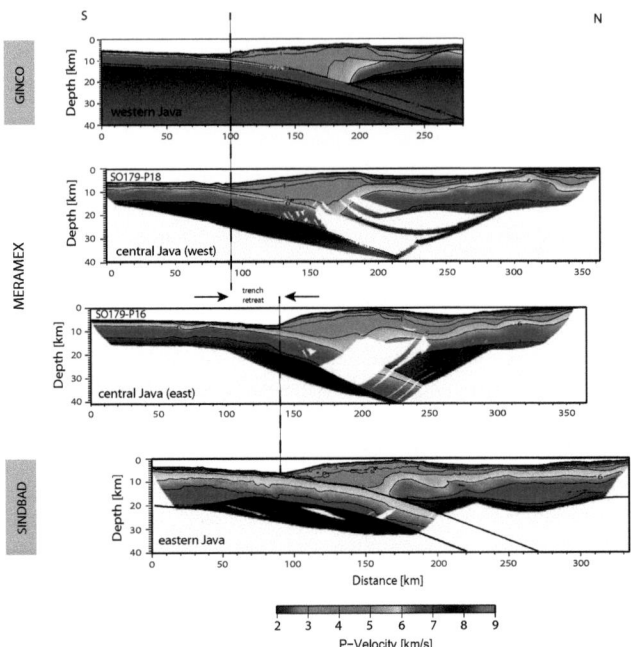

Figure 5.2: Velocity-depth distribution of all Java profiles. At the top the GINCO profile (Kopp et al., 2001), in the middle the central Java profiles of this study and at the bottom the eastern Java profile, investigated by Shulgin et al. (2009).

lithosphere at the south American trench 35° south off central Chile. The O'Higgins seamount group is located on the crest of the Juan Fernandez Ridge, where the percentage of serpentinisation is estimated with 13 % (Kopp et al., 2004). Low upper mantle velocities has been found at a number of hot spot related structures (Kopp et al., 2004).

Gravity data provide further evidence of reduced mantle velocities on the western profile. The mantle of this profile displays densities of 3.32 g/cm^3 at the Moho with increasing densities to 3.35 g/cm^3 at depth, compared to higher mantle densities of 3.37 g/cm^3 of the eastern profile. Higher densities correspond to higher P-wave velocities (Ludwig et al., 1970). The western Java (GINCO) profile shows also mantle densities of 3.37 g/cm^3 (Kopp and Kukowski, 2003). Thus mineral alteration will modify seismic velocities and lower the density compared to normal lithosphere.
Variations in the thickness and undulating velocities in the crust are common for magmatically over-

5.1. OCEANIC LITHOSPHERE AND THE CHRISTMAS ISLAND SEAMOUNT PROVINCE   89

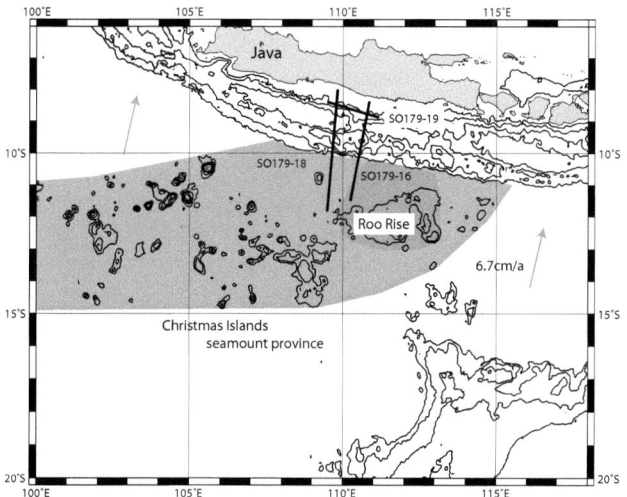

Figure 5.3: A broad band of seamounts incipiently subducts south of central Java. The Christmas Island Seamount Province (CHRISP) is a submarine volcanic province oriented in E-W direction. Contour lines range from 4500 m to 1000 m below the water surface with a contour interval of 1000 m. The Roo Rise is indicated by the gray shaded area.

printed oceanic crust. This can be associated with past, and now inactive, phases of localized increased magmatic activity (Kopp et al., 2004). Reduced mantle velocities may also be associated with maqma intrusions at the base of the crust representing magma prevented from penetrating the crust and reaching the seafloor surface during a period of weakened hot spot activity (Kopp et al., 2004). Remnants of mafic rocks from an incomplete separation of mafic and and ultramafic material have been proposed to be the cause of reduced mantle velocities (≥ 7.6 km/s) underneath Cocos ridge (Walther, 2003). Figure 5.2 reveals an strong magmatically influenced oceanic lithosphere at eastern Java (SINDBAD). The crustal thickness is dramatically increased at the outer rise and the mantle velocities indicate alteration and hydration, which supports the MERAMEX results. In contrast the western Java profile displays 'normal' oceanic mantle velocities of 8.0 km/s. The elevated P-wave velocities of 7.4 km/s in the lower crust on profile SO179-P16 have been mapped for numerous hot spot related volcanic structures and are commonly attributed to underplating of melt beneath the crust (Kopp et al., 2004).

The prevailing tectonic feature off central Java is the trench retreat of approximately 60 km, resulting from the subduction of the elevated oceanic basement relief of the Roo Rise (Fig. 5.3). The Roo Rise is 2 – 2.5 km higher than the surrounding seafloor and exerts a significant influence on the subduction off central Java. Oceanic plateaus, often in the vicinity of a hot spot, affect the thickness of the oceanic crust.

The joint inversion of PmP phases of stations OBS41 (Fig.'s 3.16, 3.17) and OBS42 (Fig.'s 3.14, 3.15) and refracted first arrival traveltime tomography (chapter 3) show a thickened subducting oceanic crust (Fig. 5.1). Testing the reliability of the crustal thickness, different Moho reflectors with variable depths and shapes are introduced to the joint tomography (Fig. 3.12) converged at a Moho depth of 9 km ± 0.5 km bsf for the western profile (SO179-P18) and a depth of 10 ± 0.5 km bsf (eastern profile SO179-P16). The critical distance of PmP reflections fits the data (Fig. 3.13), confirming the velocity model and the position of the Moho. Due to the lack of PmP phases, the Moho depth of the western profile could only be resolved at the seaward part of the trench (Fig. 5.1). This results in a 2.5 - 3.5 km thicker oceanic crust compared to a 7.4 km thick oceanic crust of the southern Sumatra- and Sunda Straight (Kopp et al., 2002) transects. The western profile lies in the transition zone of 'normal' to thickened crust whereas the eastern profile lies at the termination of the Roo Rise and follows the trend of the Roo Rise, broadening eastwards (Fig. 1.2). The crustal thickness is increased where the oceanic crust has been altered by the emplacement of the Roo Rise.Shulgin et al. (2009) observed a dramatically increased crustal thickness of up to 18 km at eastern Java. This confirms the incipient subduction of the Roo Rise at central Java. Crustal thickening mainly occurs in the lower crust and seismic and gravity data confirm the presence of a crustal root here (Shulgin et al., 2009) as postulated by Newcomb and McCann (1987) to explain the absence of a correlated gravity anomaly. These results are confirmed by numerical models, which predict crustal thickening to be concentrated in the gabbroic/basaltic layers (van Hunen et al., 2002).

Graindorge et al. (2004) recognized a comparable 14 km overthickened oceanic crust at the Carnegie Ridge. The Carnegie Ridge is a 300-km wide oceanic plateau at the Ecuador margin, which rises 1500 m above the seafloor. Its origin is the Neogene interaction between Galapagos hotspot and the Cocos-Nazca spreading center. The Carnegie Ridge underthrusts volcano-oceanic accreted terranes of the Ecuadorian margin (Hughes and Pilatasing, 2002).

Figure 5.3 shows the contour lines from 4500 m rising to 1000 m depth with a contour interval of 1000 m. These indicate large topographic highs, which rise at least 4500 m above the surrounding seafloor. Small scale mounts are excluded at a depth range of 3500 m. At the SO199 cruise in autumn 2008 the Christmas Island Seamount Province (CHRISP) was recently investigated with magnetic, swath mapping data. The seafloor was dredged to collect biological and rock samples. The CHRISP is a submarine volcanic province of unknown origin with an area of 1800 x 600 km and is oriented in E-W direction which subducts south of central Java (Fig. 5.3). The eastern segment of the CHRISP features the oceanic Roo Rise. This results in accelerated subduction erosion due to the interaction at the trench (Clift and Vannucchi, 2004). The swath mapping on the SO199 cruise SW of the MERAMEX investigation area revealed guyot-type seamounts which represent former island volcanoes. Guyot-like seamounts are characterized by circular, steep-sided bases and relatively flat tops (Werner et al., 2009). The CHRISP shows a broad variety of volcanic structures from small, isolated volcanic cones to huge plateau-like structures. In the eastern part of CHRISP, close to the central Java trench, SO199 investigated 12 major seamounts which rise from 5000 m to 2000 m below sea level. Dredged rock samples from the plateau retrieved strongly altered olivine phyric lava fragments with Mn-crusts (Werner et al., 2009). The dredging of rock samples at these seamounts yielded volcanic rocks, including highly porphyric lavas and a wide range of volcaniclastic rocks without a clear formation time trend (Werner et al., 2009).

5.2 Interplate processes

5.2.1 Segmentation of the forearc high

Different styles of subduction processes prevail along the Sunda margin (Kopp et al. (2002); Kopp et al. (2006)). The Sunda Arc has traditionally been seen as an accretionary system (Clift and Vannucchi, 2004), however, more detailed investigations such as MERAMEX are necessary to better characterize the subduction style for each segment of the Sunda margin. At the Java margin tectonic erosion with a small frontal prism as proposed by (Kopp et al., 2006), instead of a accretionary wedge, could be validated on the basis of the seismic wide-angle data presented here. Important factors for tectonic erosion are fast convergence rates of $\geq$ 6.3 cm/a (Clift and Vannucchi, 2004), which do not allow the deposition of large amounts of sediment in the trench because time considerations show that only a thin sedimentary layer may develop (Lallemand, 1994). At central Java the convergence rate of 6.7 cm/a (Tregoning et al., 1994) corresponds to an erosive regime. A rough plate interface controls erosive processes. If topographic highs interact with the frontal prism, mass wasting and slumping occurs.

Figure 5.4 shows a qualitative comparison of the outer wedge and inner wedge volume. It reveals a decreasing volume of the outer wedge from western to eastern Java and an increasing taper angle. The western Java transect (GINCO) was investigated on the basis of a combined reflection-refraction forward model of Kopp et al. (2009). The MERAMEX- and SINDBAD models are based on the tomographic inversion results (this study and Shulgin et al. (2010)).

The internal architecture of the forearc is characterized by multiple kinematic boundaries between the trench and the Java continental slope. The deformation front marks the transition from the trench to the frontal prism. The frontal prism forms the apex of the upper plate wedge and consists of frontally accreted, fluid-rich and thus mechanically weak material (von Huene, 2008). The frontal prism transitions into the Neogene accretionary prism, which rapidly increases in thickness. A pronounced out of sequence thrust or backstop thrust separates the Neogene prism from the older, more consolidated material of the Paleogene prism. The Paleaogene prism forms the core of the large bivergent wedge (Kopp et al., 2009). It is composed of fossil accreted material, which shows a positive landward gradient in the rate of lithification. Any nonuniform phase of accretion will cause a discontinuity in the lithification and thus a contrast in strength sufficient to classify the landward portion as backstop to the unconsolidated ocean basin deposits and turbidites accreted at the deformation front (Kopp and Kukowski, 2003).

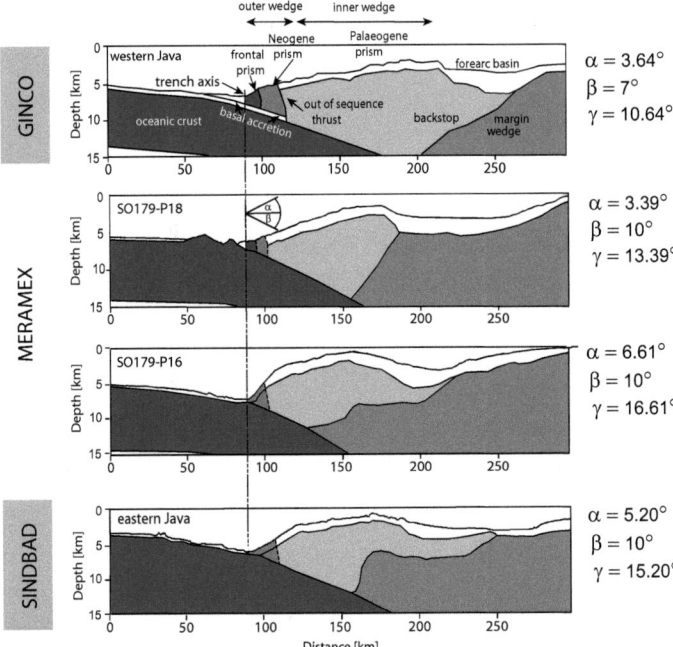

Figure 5.4: MERAMEX profiles compared with GINCO and SINDBAD. Qualitative volumes of the active outer wedge. α corresponds to the forearc slope angle, which is measured over ≥ 50 km to avoid local anomalies. β is the dip angle of the downgoing plate. The taper angle γ (the sum of α and β) increases from Sumatra to central Java. The forearc high is spearated in different kinematic boundaries. The outer wedge includes the frontal prism (dark blue) and the active Neogene prism (light blue). It is bounded by an out of sequence thrust to the Paleogene inner wegde.

The outer wedge (Fig. 5.4) includes the frontal prism and the active Neogene accretionary prism. Offshore western Java, frontal sediment accretion dominates and approximately 2/3 of the trench sediment sequence is incorporated into an imbricate thrust belt (Schlueter et al., 2002). The volume of material subducting beyond the frontal accretionary prism ranges from 500 - 1000 m per trench km here. Basal accretion likely occurs below the forearc high, contributing to the evolution and uplift of a ≥ 100 km wide bivergent accretionary wegde (Fuller et al. (2006) and Kopp et al. (2009)). To

5.2. INTERPLATE PROCESSES

the east offshore central Java, the transition from sediment accretion to tectonic erosion occurs over a distance off less than 100 km. Here, the trench is devoid of sediments except for isolated sediment ponds (Masson et al., 1990). A complex canyon system traverses the continental slope and supplies material to the Java and Lombok forearc basins. Sediment discharged from Java and the Lesser Sunda islands does not reach the trench, and is trapped in the forearc basins.

The stations OBH36 of profile SO179-P16 and OBH47 of profile SO179-P18 document the increase of seismic velocities from the outer wedge (phase Psed) to the inner wedge (phase Pg forearc). The outer wedge of the eastern profile SO179-P16 is almost completely eroded and a frontal prism is not present here (Fig. 5.4). The transition to the inner wedge is based on an estimation, because a distinct boundary could not be resolved in the seismic sections (dashed line in Fig. 5.4). This trend of erosive subduction extends to eastern Java. A less sediment (accretionary) input and the absence of basal accretion determines a decreasing outer wedge (Fig. 5.4). The erosive subduction is documented by a strongly increasing forearc slope angle α from western Java of 3.6° to more than 6° off central Java. The slope angle is measured over a distance of $\geq$ 50 km to avoid local anomalies (Clift and Vannucchi, 2004). The increasing dip angle β from western Java (7°) to eastern Java (10°) and the increasing frontal slope angle α documents the increasing taper angle γ, which will be discussed in the next chapter.

The investigated models reveal different backstop geometries between the two MERAMEX profiles. This results in varying forearc basin structures as explained in chapter 5.2.4. The forearc is defined as a part of the upper plate seaward of the volcanic arc at a subduction zone. A wide range of material exist here from porous sediments to well-lithified rocks (Byrne et al., 1993).

The backstop is defined as a region within a forearc that deforms only little and consist of stronger material, which can not be closer specified. This material can consist of igneous rocks or accreted and lithified sediments. The main attribute of a backstop is its ability to support larger deviatoric stresses and has a higher shear strength than the forearc material above or the sediment lying farther trenchward. The dominant force affecting the wedge is a resistive drag along its base and the vertically integrated strength of the upper plate increases with distance from the deformation front. The requirements for increasing strength is an increase in the thickness of the overriding plate (tapered or wedge shape).

However, the change of the subduction style from western to central Java provides only a very limited active outer wedge and a frontal prism could not be identified at SO179-P16 and the neighboring SINDBAD profile.

5.2.2 Critical taper analysis

The theory of critical Coulomb wedges based on Davis et al. (1983) and Dahlen (1990) explains the morphology of accretionary prisms. Noncohesive Coulomb material within the wedge is deformed by traction along the base as imposed by the subduction of the oceanic crust. The friction along the decollement at the base of the wedge achieves a balance to the deformation within the wedge, until a critical taper is obtained, after which it slides.

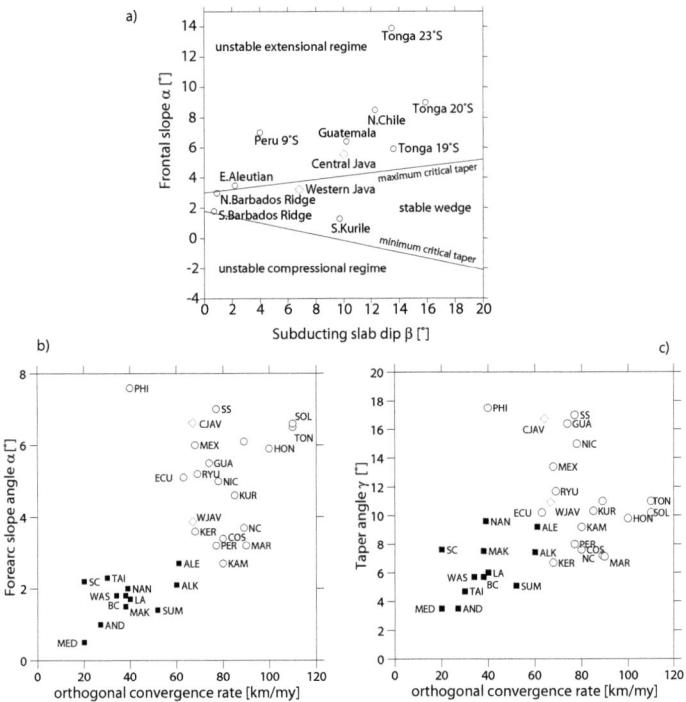

Figure 5.5: a) This graph outlines the stability field of tapered wedges. The taper of central Java lies in the unstable extensional regime, western Java in the stable wedge (red diamonds). The function of the forearc slope angle α (b) and taper angle γ (c) over the orthogonal convergence rate of the oceanic crust. If the convergence rate exceeds 6,3 cm/a the subduction style is erosive (Clift and Vannucchi, 2004). Black squares indicating accretionary and white circles erosive subduction systems.

The steepness of the upper plate is an important aspect concerning the slope stability in the wedge. The slope angle α of the upper plate as a function of the dip angle β of the downgoing plate comprise the critical taper angle γ (Lallemand, 1994). An undercritical wedge will only be deformed internally and will not slide while an overcritical wedge will slide without compressional deformation, which results in slumping and erosion (Lallemand, 1994). The taper of the wedge depends on the coefficients of friction within the material (high friction results in small tapers) and of the detachement (high friction results in large tapers). Therefore the critical taper angle function yields the stability field of the margin, dependent on material values for internal and basal friction and pore fluid pressure ratio. The angles above the maximum critical taper (Fig. 5.5) are classified as erosional style in the unstable extensional regime. The critical taper angle γ of central Java lies in the unstable extensional regime, while the western Java part lies in the accretionary stable wedge (Clift and Vannucchi, 2004). The diagrams in Figure 5.5 displays the relationships between the orthogonal convergence rate and the forearc slope angle α and the taper angle γ over wavelengths of ≥ 50 km to avoid local anomalies respectively. With an orthogonal convergence rate of 6.7 cm/y and a forearc slope angle of more than $6°$ the central Java subduction can clearly be classified as an erosive subduction. This is not the case in western Java, which has a less steep slope angle of $3.64°$ and is more typical for accretionary systems.

5.2.3 Seamount subduction and the uplifted forearc high

The eastern profile SO179-P16 (Fig. 5.1) reveals a subducted seamount at a depth of 15 km around profile-km 190. The inner wedge offshore Java is characterized by velocities generally not exceeding 5.0 km/s (Fig. 5.2). The subducted relief is inferred from the higher velocities (≥ 5.4 km/s) at the base of the inner wedge retrieved along profile SO179-P16. OBH30 (Fig.'s 3.27 and 3.28) covers the entire forearc and records the internal structure of the inner wedge and the subducting slab. Imaging, however, is intricate due to the severe deformation in this domain. In addition to the deformation of the overriding plate, the seamount itself experiences faulting and possible rupture. Baba et al. (2001) investigated the stress field associated with seamount subduction and concluded that shear failure and fracturing or dismemberment of subducting seamounts occur. This will affect seismic velocities and limit the velocity contrast between the inner wedge and the subducted seamount. In addition gaps in raycoverage along SO179-P16 inhibit the imaging (Fig. 5.2).

The presence of a seamount is further supported by a number of very distinct surface effects that document the dynamic influence of seamount subduction on the forearc morphology. These effects are associated with the subduction of moderate sized features (Dominguez et al., 2000) and include local surface uplift enhanced by compression, topographic perturbation of the lower slope, intensification of subduction erosion, and landward trench displacement.

The surface effect of seamount subduction and the subsequent evolution of the forearc is imaged by the high resolution bathymetry dataset, recorded on the SO179 cruise (Fig.'s 5.6 and 5.7). Especially the deformation of the lower slope is revealed well in the absence of a thick sediment apron. The topographic perturbations resulting from subduction of oceanic relief depend on the size and structure of the subducted feature and on the nature of the overriding plate. Subducting seamounts cause local uplift and penetrative fracturing of the overriding plate (von Huene et al., 2000) and leads to large scale seafloor failures which results in several debris avalanches at the central Java trench axis (Fig. 5.7).

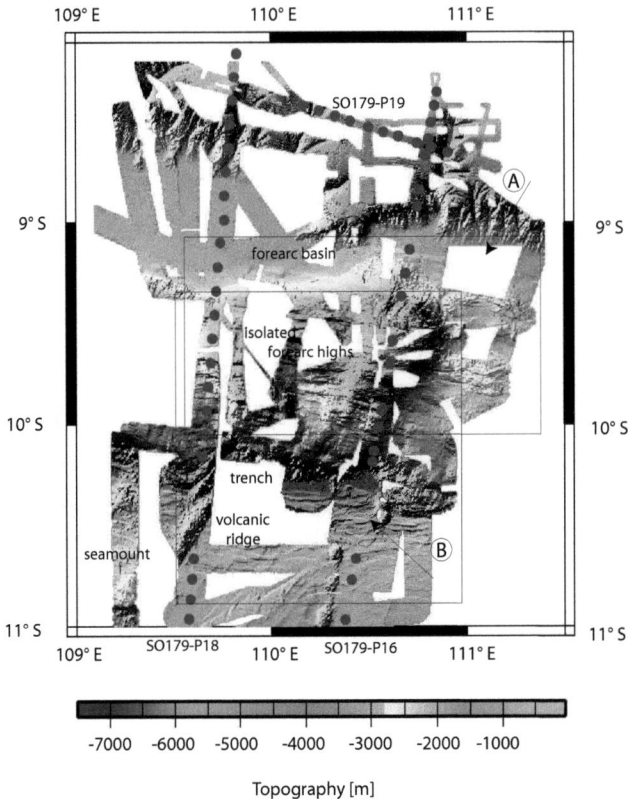

Figure 5.6: High resolution swath mapping of the investigation area. Two squares (A, dashed border and B, solid border) showing the area of perspective view at the right hand side. Station locations marked by red circles.

5.2. INTERPLATE PROCESSES

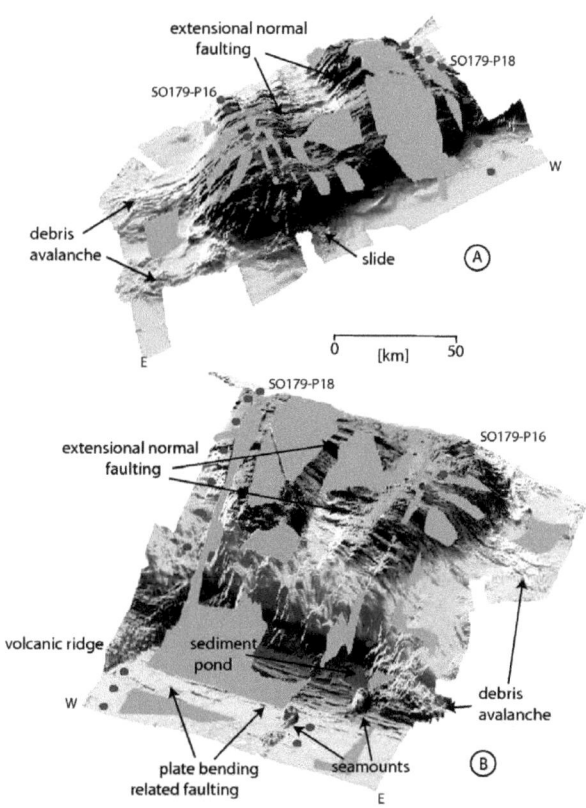

Figure 5.7: High resolution swath mapping of the investigation area in a perspective view. A) Displays the forearc basin at the bottom and the forearc high with a view from a landward NE - SW direction. B) View from a seaside SE - NW direction with the oceanic plate, the trench and the forearc high. Station locations marked by red circles.

Seamount subduction has been investigated at erosive margins (e.g. von Huene et al. (2000)) where the seamounts leave pronounced re-entrant grooves as they plough through the small frontal prism before being subducted beneath the continental framework rock (von Huene, 2008). Comparable embayments are not as distinct offshore Java (Fig. 5.7), where the accretionary material behaves more plastically. Frontal erosion has sculpted the lower slope off central Java and is associated with a northward retreat of the deformation front up to 60 km (Kopp et al., 2006). This segment of the Java margin shows very high surface slope values at the lower slope of the overriding plate of almost 7° (Fig. 5.4) and therefore bringing the taper into the unstable domain. This results from compression of the forearc, which primarily causes deformation and uplift of the thin leading-edge of the forearc (Taylor et al., 2005). The unstable frontal prism is marked by re-entrant scars correlated with sediment-ponds (Fig. 5.7), by mass failure and extensional normal faulting. Erosive processes are enhanced by the lack of sediment in the trench and the pronounced horst-and-graben structure in the trench where the plate bends underneath the forearc. The bathymetry data reveal the incipient subduction of a small volcanic ridge (Fig. 5.7) currently positioned in the trench. Larger topographic features on the oceanic plate are resolved by the local- and global bathymetry like a seamount of 70 km in diameter and 3 km of height (Fig.'s 4.2 and 5.6) and the CHRISP province (Fig. 5.3). Topographic perturbations resulting from seamount subduction within the outer wedge are transient and the prism will heal after the relief is subducted to greater depth. The observed uplift on profile SO179-P16 is inferred to be caused by the impingement of oceanic basement relief and the associated compressional deformation. A trench perpendicular compressional force is applied on the forearc by the relatively buoyant and thick subducting Roo Rise and its volcanic seamounts. This effect has also been reported for other margins, e.g. the Ryukyu margin (Font and Lallemand, 2009) or Hikurangi margin (Litchfield et al., 2007). The uplift results from isostatic adjustment and is enhanced by crustal shortening of the upper plate. The trench perpendicular compression leads to surface elevation of the forearc high, which greatly exceeds the original height of the seamount, as predicted by numerical modeling (Gerya et al., 2009). Not all the uplift can be due to the volume of the seamount but that a part of it must be caused by landward displacement of wedge material. The observed surface uplift of 1 km on the eastern profile SO179-P16 correlates with the supposed position of the seamount. The uplift is generated by crustal shortening and thickening of the upper plate over a locked segment of the subduction thrust (Taylor et al., 2005). Backthrusting of the forearc basin onto the forearc basin results from the strong compression of the entire segment and partly accommodates forearc convergence (e.g. Taylor et al. (1995)).

The transition from the active outer wedge to the inner wedge occurs along a distinct zone, where a splay fault system offsets the seafloor. This forethrust builds up the inner part of a bivergent wedge, and can reach the downgoing oceanic plate. The bathymetry (Fig. 5.7) shows a slide possibly triggered by compressional stresses along the forethrust fault. However the bathymetry (Fig. 5.7) shows a characteristic fault network for a subducted seamount. Seaward dipping back thrusts and

5.2. INTERPLATE PROCESSES

conjugated strike-slip faults, related to the indentation of the margin, propagate landward as the seamount subducts. Magnetic data, collected by the Federal Institute of Geosciences and Natural Resources (BGR) on the SO179 cruise, show a number of isolated short-wavelength positive anomalies in the forearc high. These coincide with bathymetric and gravimetric highs and indicate source rocks at shallow depths (Kopp and Flueh, 2004). These rocks indicate further evidence of subducted seamounts.

5.2.4 Forearc basin and submarine landslides

The Java basin is expressed by an elongated, 500 km long subsiding belt with an average water depth of 3500 km. Its evolution is governed by the accretion-driven uplift of the forearc high, which forms the barrier to the trench and abyssal plain and by tectonically induced subsidence forming a rapidly filled depression (Susilohadi et al., 2005). The sediment thickness reaches 4 km, decreasing towards the basin frindge. The oldest sequences in the basin were deposited in the Middle Eocene to late Oligocene. The sediment supply increased during the late Middle Miocene with the rising volcanic activity of the arc (Susilohadi et al., 2005).

The forearc basin is underlain by a unit characterized by seismic velocities rapidly increasing from 5.5 km/s to values larger than 6 km/s (Fig. 5.2). OBH53 (Fig.'s 3.23 and 3.24) is situated on the northern rim of the forearc high on the western profile SO179-P18 and records phases through the outer and inner wedge as well as through the forearc crust (phase Pg margin) and the mantle. The forearc basement below the forearc basin shows an ophiolithic character. The basement is exposed in western Java, where outcrops of peridotites, gabbros, pillow basalts and serpentines are observed (Susilohadi et al., 2005).

The landward limit of the inner wedge is terminated by the margin wedge (Fig. 5.4), which acts as a static backstop (Kopp and Kukowski, 2003). Following the 5 km/s iso-line (Fig. 3.8), the toe of the backstop is situated below the crest of the forearc high. On the western profile SO179-P18, the static backstop casts a stress shadow over the area itself, allowing the presence of a forearc basin which is not experiencing any deformation as sediments are deposited within it (Byrne et al., 1993). A largely undeformed forearc basin with deposited sediments growths in the stress shadow over the static backstop. Byrne et al. (1993) calculated the stress and displacement fields within forearcs for a variety of backstop models using the finite element method and a small scale sandbox model. Numerical models show that in response to applied basal shear all models experience internal deformation involving shortening and uplift. Above the toe of the backstop the direction of maximum principle stress rotates from a trenchward dip within the accretionary wedge to a steep landward dip above the backstop. Region of failure with thrusts forms an inner deformation belt that grows with uplift of the outer arc high. The largest deviatoric stresses occur within the backstop near its toe, but greater material strength supports the larger stresses without large strains. This protected region provides a site for the formation of an undeformed forearc basin. The effects are independent of the rheology (used models with elastic, viscous and Mohr-Coulomb behavior respectively).

A margin can change from one type of backstop to another. Possible mechanisms include rearranging of the lithification front either by a major change in the rate or type of sediment accretion, or the collision of a major bathymetric feature. The velocity contrast between the inner wedge to the margin wedge is much higher on the western profile SO179-P18 (0.6 km/s) compared to profile SO179-P16 (0.1 - 0.3 km/s). Owing to the fact, that this region is less resolved in the tomography, it is supposed that the subducted and dismembered seamount, situated at the toe of the backstop on the eastern profile SO179-P16, influences the strength of the backstop, which results in modified porosity contrasts. The top of the backstop is approximately 3 - 4 km deeper below the forearc crest and is located closer to the trench. As the outer high is uplifted on profile SO179-P16, the seaward part of the forearc basin experiences some uplift and deformation, resulting in landward tilting and pinching out of the older strata near the outer high. Backthrusting of the forearc high onto the forearc basin results from the strong compression of the entire segment and partly accommodates forearc convergence (Taylor et al., 1995).

The uplifted forearc high determines steep slopes of the frontal prism and the slope to the forearc basin. The main trigger for slope failures are oversteepened slopes due to the subduction relief or near a thrust surface in the frontal imbricate thrust fan and earthquakes. Even a small earthquake can trigger a landslide at slopes close to failure (Brune et al., 2009). Weather a submarine mass movement generates significant wave amplitudes for a tsunamigenic event depends on its volume, water depth, shape and its velocity profile (Brune et al., 2009). Brune et al. (2010) identified landslides with a volume of 0.1 - 20 km^3 at twelve different locations along the Sunda margin. The largest slides were found at the eastern Java trench with volumes of 1 - 20 km^3. The main reason for the large slide volumes are the locally oversteepened slopes due to the erosive subduction regime along the eastern Java margin. Despite the large sediment supply to the Sumatran accretionary prism no large landslide could be correlated with the 2004 Sumatra earthquake (Brune et al., 2009). Four slides at the eastern Java margin off Lombok to the Sumba island with their volumes up to 20 km^3 are located close to the 1977 Sumba earthquake (Mw = 8.3) and could have triggered the devastating tsunami (Brune et al., 2009).

Figure 5.7,A shows a slide with a volume of approximately 3 km^3 located in the forearc basin, 120 km off the coast of Java, which was discussed by Brune et al. (2009). Brune et al. (2009) calculated numerical models to estimate the generation, propagation and run-up height of the tsunami. Despite the significant slide volume it would not generate a large tsunami. The estimated run-up height would not exceed 1 m, and is comparatively small to those of 6 m generated by the large slides off Sumba island. Figure 5.7,B displays a large debris avalanche of approximately 25 km width at the toe of the frontal slope. The failure mechanism seems to be an oversteepened slope of the uplifted forearc high. It is unknown weather this mass wasting triggered a tsunami or not. However, the bathymetric dataset is not complete and could reveal other sites in the forearc basin which have the potential to trigger a more catastrophic tsunami.

5.3 Seismicity and megathrust earthquake potential

Based on the data recorded with the temporary local onshore seismological network (Fig. 1.7), the German Research Center for Geosciences (GFZ) investigated the distribution of approximately 500 local earthquakes with magnitudes ≤ 4. Figure 5.8 displays these epicenters, color coded by depth distribution. Along a NS striking slice (AB), starting on the outer rise and ending NE of the Merapi volcano, all earthquakes between 104°E and 118°E are projected along this slice. The epicenters are concentrated south of the local network off the Javanese coast with a major depth distribution between 35 to 150 km.

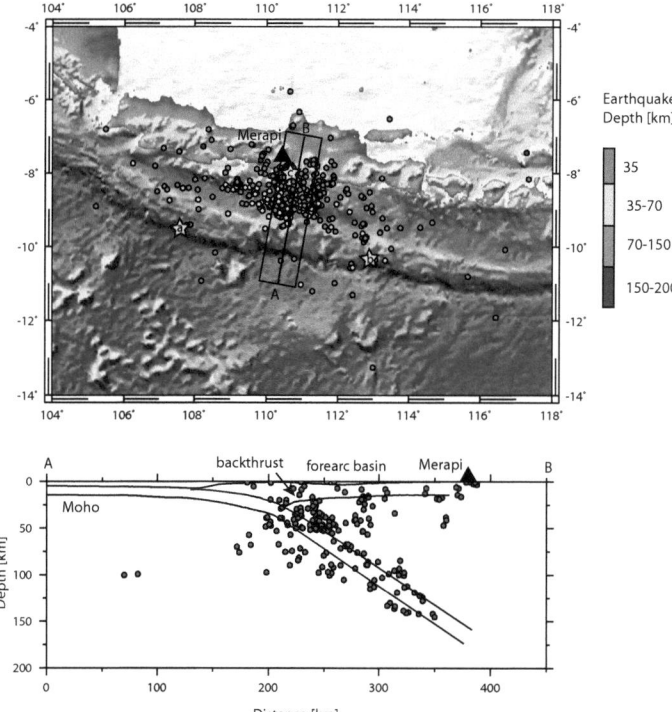

Figure 5.8: Seismicity of 500 events during the MERAMEX experiment investigated by the German Research Center for Geosciences (GFZ). The depth distribution is color coded. The yellow stars indicate earthquakes described in the text. a: 2006 July 17, Mw=7.7; b: 1994 June 02, Mw=7.8; c: 2006 May 26, Mw=6.4. 220 earthquake hypocenters are projected in a corridor of 50 km (dashed box) onto the landward extended line SO179-P16 (AB).

Earthquake events in the depth range of 40 to 150 km show a seismic double zone which could also be confirmed by Koulakov et al. (2007). Similar double seismic zones have been observed at other subduction zones (e.g. Nakajima et al. (2001)). The intermediate depth events are presumed to be related to phase transition of blueshist to eclogite (Peacock, 2001). The double seismic zone is supposed to be related to the isotherms in the subducted oceanic slab (Koulakov et al., 2007).

The Wadati-Benioff zone, which images the downgoing slab with the seismicity, shows a variable dip angle. The slab appears to be almost horizontal with a shallow dip-angle of 10° down to a depth of approximately 50 km. This could also be verified by the joint reflection and refraction tomography (chapter 3) with PmP phases. In greater depths the dip angle increases to 45°. The increasing dip angle of the subducting plate (Fig. 5.4) can also be related to the increasing plate age along the Sunda margin. The 135 Ma (Moore et al., 1980) old plate segment at the Java trench determines lower lithospheric buoyancies and therefore a steeper dip anlge (Brune et al., 2010). Seismogenic behavior is governed by changes in the size (width) of the seismogenic interplate coupling zone. The release of an earthquake is proportional to the size of the fault zone, a narrow seismogenic zone would only promote small earthquakes. The central Java segment of the subduction zone would provide a limited seismic rupture due to a narrow coupling zone determined by a shallow mantle. This leads to a narrower coupling zone between the subducting and the overriding plate, which results in the absence of huge seismic events (Burbidge et al., 2008).

The shallow earthquakes in the forearc crust are mainly aligned along the backthrust (Fig. 5.8). The potential activation of the corresponding splay faults or out-of sequence thrusts during the co-seismic phase plays an important role for the tsunami generation. Splay faults connect to the megathrust at depth and dip steeply to the surface, as imaged offshore western Java (Kopp et al., 2009). The low-angle slip of the megathrust will be potentially transferred to a higher angle, which may greatly enhance seafloor displacement (Tanioka and Satake, 1996).

No significant earthquake activity was observed in the seismogenic zone between the outer wedge and the backthrust (Fig 5.8). The reason for the quiescence of earthquakes in the seismogenic zone could be a permanent steady creeping of the weakly coupled upper and lower plates. Therefore not enough strain can be established for larger earthquakes.

A cluster of earthquakes is present in the mantle wedge. As inferred by thermal models of the NE Japan subduction zone, they occur in the coldest part of the forearc mantle wedge (Uchida et al., 2010), just below the forearc Moho. The forearc Moho reaches a minimum depth of 16 km in central Java (Fig. 5.1), intersecting the downgoing plate at ≤ 20 km. The existence of a shallow mantle wedge was already proposed based on wide-angle seismic data (Kopp et al., 2002). The forearc Moho is shallower than the observed downdip extend of the seismogenic zone. This does not support the hypothesis that the Moho is its downdip limit. Klingelhoefer et al. (2010) observed aftershocks and located the main shock of the 2004 Sumatra earthquake in the forearc mantle. The rupture occurred along the interface between the oceanic crust of the downgoing plate and the forearc mantle of the

5.3. SEISMICITY AND MEGATHRUST EARTHQUAKE POTENTIAL

upper plate. Hino et al. (2000) observed the Sanriku-Oki earthquake with a magnitude of 7.7 which ruptured in the mantle wedge beyond the downdip limit of the seismogenic zone. Uchida et al. (2010) called these events supraslab earthquakes and observed these clusters down to a depth of about 50 km.

Fluid-pressure effects of a dewatering downgoing slab into the cold forearc mantle would reduce the effective normal stress and promote brittle fracture and unstable frictional sliding at depth. Free water as a fluid phase is expected to be unstable in the presence of peridotite at low temperatures, producing serpentine (O'Hanley, 1996). Also antigorite, the high-pressure form of serpentine, exhibits velocity-strengthening behavior in the laboratory and therefore exhibits frictional sliding (Reinen et al., 1994). Thus earthquakes are not expected in serpentinised mantle or along serpentinised fault zones in the forearc mantle.

Uchida et al. (2010) infers that the supraslab earthquake clusters may represent a seismological expression of underplating occurring by the detachment of seamounts. At depths $\geq$ 30 km and temperatures of 100 - 200° the altered minerals of seamounts should dehydrate (Oleskevich et al., 1999) and have the potential to create a high fluid pressure at the base of seamounts. Under these conditions seamounts can detach more easily than at shallow depths. Internal structures, such as rifts, slump and landslide block boundaries, and normal fault zones generated by seamount bending at trenches, should also be altered. The release of water by dehydration could lead to a tectonic dismemberment of seamounts into fragments (Uchida et al., 2010). This study reveals a dismembered subducted seamount at profile SO179-P16, which is located at approximately 20 km depth at profile-km 200 (Fig. 5.1). It can be supposed that formerly subducted seamounts cause the earthquake clusters, which have been fragmented and mechanically dismembered during subduction along internal planes of weakness. If we assume a series of subducted seamounts along the eastern profile, the process of detaching and dismembering should be sequential. Accreting seamounts in the forearc region are another possible explanation for the uplifted forearc high at central Java, as suggested by Uchida et al. (2010) for the uplift in the Tohoku forearc. However, the observation of earthquakes in the forearc mantle does not support the hypothesis, that the seismogenic zone is only controlled by the dimensions of the coupling zone between the crust of the overriding plate and the subducting plate.

Not only the size of the interplate coupling zone plays a role in generating large earthquakes. Also the subducted basement relief is acting as asperities along strike will also resist co-seismic slip. An asperity is an area with locally increased friction and exhibits a reduced amount of interseismic aseismic slip relative to the surrounding regions. Hence, it will slip in an increased amount during an earthquake (Das and Watts, 2009). Subducting seamounts may prevent earthquakes from propagating through an area. Such a barrier would have either very high friction to prevent earthquake from propagating or low friction (Bilek and Engdahl, 2007). Stress during the interseismic period is not stored in case of low friction and is released aseismically or in small earthquakes. The absence of stress results in prevention of earthquake propagation (Das and Watts, 2009). Mochizuki et al.

(2008) observed low friction along the plate interface over a subducted seamount and in the wake of its subduction in the Japan trench. Stress is concentrated at the subduction front where large earthquakes with a broad rupture area can be initiated (Mochizuki et al., 2008). Thus subducted bathymetric relief acts to both nucleate seismic rupture and also to limit lateral rupture propagation. The heterogeneous and complex interface geometry is the main reason for the absence of large megathrust earthquakes offshore Java, while smaller earthquakes frequently occur. In contrast, off Sumatra, the megathrust is not as commonly perturbed by subducting relief as offshore Java. Another important aspect for rupture propagation is the amount of sediment in the décollement zone. Material transported in the subduction channel favors rupture propagation. Fluid-rich sediments reduce the effective normal stress due to a elevated pore pressure and determine a weak coupling zone (von Huene et al., 2004). In the Antilles, 700 - 1200 m of sediment in the décollement zone would potentially support large rupture (Becel et al. (2010); Weinzierl (2010)). In comparison the margins of Java, southern Chile or Cascadia display an oceanic crust topography, smoothed by a higher sediment supply and by weak forearc compression. Large megathrust earthquakes occur on these margins, where along strike co-seismic slip is not hindered by basement asperities while the downdip extend of the seismogenic zone is larger than off Java. Off Sumatra, a thickly sedimented trench and décollement zone filled with trench sediment is observed (Mw $\geq$8.5). Off Java, a starved trench will not smooth subducted relief.

The main reason for subduction related earthquakes is the release of stress due to a reverse slip of the upper plate along the surface of the downgoing plate. A special type of these earthquakes is called "slow" or "tsunamigenic" earthquake (Kanamori, 1972). The events from 1994 and 2006 (Fig. 5.8) were tsunami earthquakes. These events are characterized by large tsunami run-ups relative to the seismic moment, the earthquake energy is high-frequent and the rupture has a long duration (80 - 90 s for the 1994 event and 180 s for the 2006 event). These earthquakes with a similar magnitude Mw=7.8 (1994) and Mw=7.7 (2006) had locations close to the Java trench. Both earthquakes produced aftershock sequences dominantly comprised of normal faulting events. This is unusual for large reverse mechanism mainshocks (Bilek and Engdahl, 2007). Maybe related to a weak coupled plate interface, it was not capable to produce many reverse mechanism events on the interface. The oceanic plate off Java has a very rough plate interface with numerous seamounts, which influences the rupture characteristics of these earthquakes and their aftershocks. Abercrombie et al. (2001) found the highest slip in an area of a previously subducted seamount. The margin was weakly coupled outside of these subducted features. Therefore the seamounts act as asperities during large earthquakes. The rest of the interface is decoupled, hence no reverse mechanism aftershocks occur in the outer rise due to the extension of the subducting plate in response to the downdip slip at the locked seamount patch.

A heterogeneous coupling zone, with low sediment coverage and a rough downgoing oceanic plate, dotted with seamounts determines a high stress environment with a strong interplate coupling (Tanioka and Satake, 1996). Seamounts underthrusting the slope region cause primarily strong aseismic defor-

5.3. SEISMICITY AND MEGATHRUST EARTHQUAKE POTENTIAL

mation of the overriding plate. Thus an enhanced interplate coupling can also increase the earthquake recurrence interval (Scholz and Small, 1997). A similar effect is exerted by voluminous subducting ridges or plateaus. The Ecuador margin is divided in two different segments. The southern margin segment, where the bulk of the Carnegie Ridge is subducting, was not effected by large earthquakes for the last century (Graindorge et al., 2004). The adjacent northern segment ruptured twice over the same period. The model of Scholz and Small (1997), developed for subducting seamounts, also appears to be valid for large and thick volcanic ridges. The buoyant Roo Rise south off central Java seems to be responsible for the accumulation of normal stress and a uplifted forearc high. Together with subducting seamounts the interplate coupling is very heterogeneous.

Whereas the highest slip of the 1994 event is concentrated in a zone of uplifted topography, the slip during the 2006 event terminates in a region west of a large segment of uplifted topography. This region of uplifted topography seems to indicate a barrier to continued slip. These slip barriers and asperities have been discussed for many earthquakes in other fault zones (e.g. Cloos (1992); Seno (2002); Robinson et al. (2006)).

The earthquake of 2006 May 26 was located in a low-velocity zone between two rigid blocks (Wagner et al., 2007). This can be interpreted as a weakened contact zone between two rigid crustal bodies, which will be discussed in the next chapter.

5.4 Comparison of the onshore and offshore tomography results

The onshore tomography results are presented in Wagner et al. (2007) and Koulakov et al. (2007).

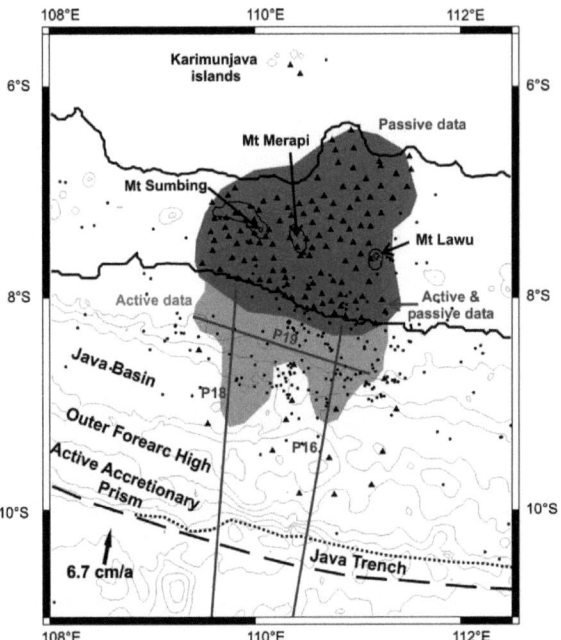

Figure 5.9: This figure is from Wagner et al. (2007). Study area with the onshore experimental setup and the tectonic features of the region. Triangles mark the temporary seismological network. Dots refer to the earthquakes used for the tomography. The red colored area in central Java marks the investigation area covered by passive data, the light blue is covered by active seismic data and the grey area marks both data sets in the uppermost 10 km depth layer.

The tomography revealed a low velocity anomaly northward of the volcanic arc. The volcanoes Sumbing, Merapi and Lawu are above a very sharp boundary between the high velocity forearc and a very strong low velocity anomaly with -30 % located northward of these volcanoes in the backarc crust. Beneath 15 km depth the negative anomaly increases in size and moves southwards and decreases in amplitude (Fig. 5.10). For these results active and passive data are used. This anomaly is suggested to correspond to multiple magma reservoirs and ascending feeder systems Wagner et al. (2007). Another negative velocity anomaly is located at the western dip line in NS direction (Fig. 5.10).

5.4. COMPARISON OF THE ONSHORE AND OFFSHORE TOMOGRAPHY RESULTS   107

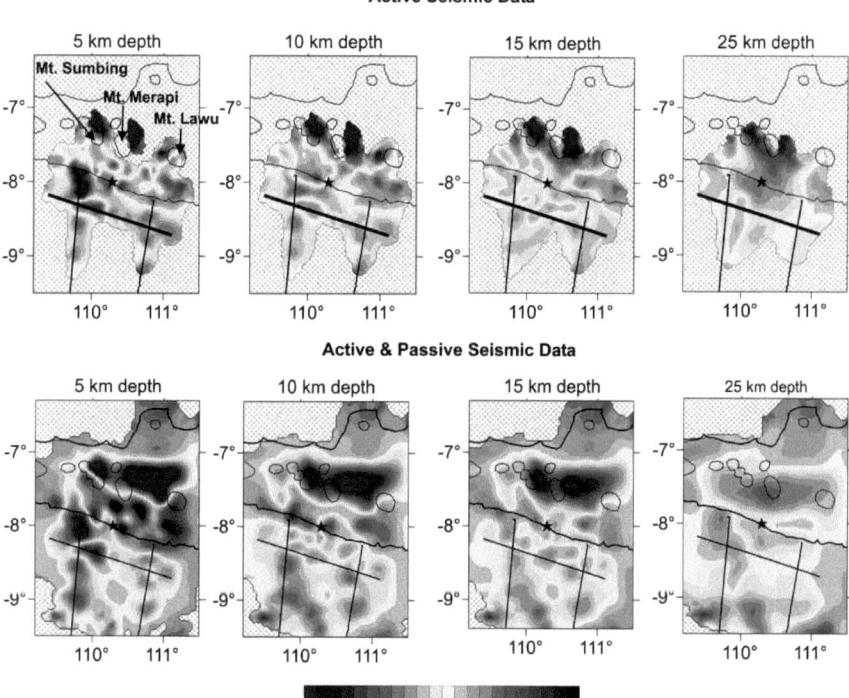

Figure 5.10: This figure is from Wagner et al. (2007). P-velocity anomalies of two datasets. Top: only the active dataset was used and is presented in horizontal slices at 5, 10 and 15 km depth. Bottom: Combined dataset of active and passive data. The coastline and the wide-angle seismic profiles are included as black lines. The star indicates the location of the earthquake from 2006 May 26.

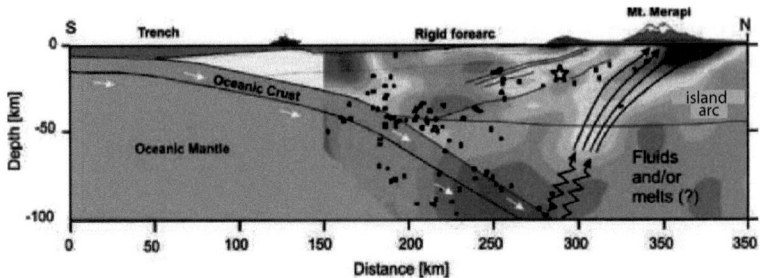

Figure 5.11: Interpretation of the velocity structure in central Java from Wagner et al. (2007). The velocity structure is limited in west to the forearc basin. Black dots display approximately 200 earthquakes, which are located in the region of the tomography. The yellow star indicates the Federal Institute of Geosciences and Natural Resources (BGR) hypocentre of the Java earthquake in 2006 May 26.

The uppermost 20 km crust in the onshore forearc consist of two high-velocity blocks. The contact zone between these two blocks is separated by an elongated low-velocity zone. The epicenter of the Java earthquake of 2006 May 26 is located at the edge of this low-velocity zone, which is interpreted as a weakened area between the two rigid forearc blocks.

The fracturing of rocks decreases the velocity in this zone in the uppermost crust. At a depth greater than 15 km, where the earthquake was located, this low velocity zone is almost invisible. This implies that it was located in the rigid crust, just below the weakened low-velocity zone, which is the most probable location for stress accumulation and rupture (Wagner et al., 2007). The high velocity anomaly from the tomographic results of Wagner et al. (2007) at the intersection of SO179-P18 and SO179-P19 corresponds with the higher velocities of the inherited basement high.The increasing dip angle of the slab would determine a northward pushing and stress accumulation in the upper plate (Wagner et al., 2007). The observed earthquake of 2006 May 26 can be the result of this mechanism (Fig. 5.11). The inclined linear anomalies in the forearc might reflect the distribution of weakened fracture zones (Wagner et al., 2007).

The volcanism in central Java is related to the subduction process. The earthquakes in the Wadati-Benioff zone around 100 km depth are related to phase transitions in the slab causing fluid release and partial melting of the oceanic crust (Wagner et al., 2007). The inclined low-velocity anomaly above 60 km depth may be attributed to partial melting. The rigid bodies in the shallow forearc are blocking the rising fluids ascending further upwards. After the ascending fluids and melts have reached the northern boundary of the forearc, they form high concentrations of gases and magma and are the source for the active volcanism (Wagner et al., 2007). The low-velocity anomalies in the uppermost 10 km are interpreted as lava and sedimentary deposits.

Chapter 6
Conclusion

The MERAMEX project revealed new insights in the subduction processes offshore central Java. The interpretation of three marine wide-angle seismic data profiles is based on a combined forward- and inverse modeling of first- and later arrival travel times, complemented by forward gravity modeling. Using the methodology of an alternating forward- and inverse modeling delivered results with a good accuracy, with the possibility to effectively introduce and test distinct model features. The geodynamic implications of this study are crucial to better understand the variation in the tectonic regime along the Java margin and are embedded in a series of studies along the margin. The results underscore the strong influence of the oceanic plateau Roo Rise and the volcanic edifices on the underthrusting plate on the subduction dynamics.

The western profile located at 110°E south of central Java reveals a hydrated, slightly thickened (9 km) oceanic crust. Alteration of the oceanic crust down to Moho depth and hydration of the upper mantle is governed by a large (70 km diameter) seamount found in the vicinity of the profile that causes a local moat structure and in conjunction with plate bending at the trench activates deeply penetrating faults that may serve as fluid pathways. Offshore central Java, a northward deflection of the trench axis by 60 km is observed in the regional bathymetry and is closed related to the subduction of the oceanic Roo Rise. This buoyant plateau with its thickened crust (10 km) is currently being underthrust beneath the forearc. The plateau is dotted by numerous seamounts, originating from the Christmas Island Seamount Province, the subduction of this high-gradient relief causes a transition from an accretion-dominated subduction system offshore western Java to an erosive regime off central Java over a distance of less than 100 km. A starved trench and frontal tectonic erosion of the forearc concur with an oversteepened inner trench wall and local mass wasting from the upper plate. The frontal prism, which is characterized by actively deforming sediments accreted from the lower and upper plate, is completely eroded on the eastern profile located at 111°E south off central Java. The geometry of the forearc is revealed based on the velocity-depth models and the density models derived in this study. Due to the lack of coincident, deep-penetrating multichannel reflection data, distinct tectonic boundaries such as the backstop thrust, cannot be revealed.

Inhomogeneities on the plate interface, such as subducted oceanic basement relief, potentially act as seismic asperities, as suggested for the 1994 Java tsunami earthquake (Abercrombie et al., 2001). The velocity-depth distribution along the eastern profile south off central Java is indicative of a subducted and dismembered seamount at 15 km depth. However, the role of such features, which are ubiquitous along the Java margin as documented by high-resolution bathymetry data, in seismogenesis may only be understood through local long-term earthquake monitoring.

One of the most important findings of this study is the existence of a shallow upper mantle. The upper plate Moho is traced at a depth of 15 - 20 km. This is in accordance with similar findings offshore western Java and off the Lesser Sunda islands. A shallow mantle wedge will limit the depth extent of the seismogenic zone and is key in the regional variation in subduction earthquakes, which reach moment magnitudes well above 8.5 offshore Sumatra, while moment magnitudes ≤ 8 are observed in megathrust events offshore Java. However, the recurrence of devastating tsunamis in the wake of subduction earthquakes offshore Java (e.g. the 1994 and 2006 Java tsunami events), documents the high potential for natural hazards offshore Indonesia's economic center. Full coverage of the high-resolution bathymetry map as well as multichannel seismic data to unequivocally identify the extent of the frontal prism and / or the existence of possible splay faults, are required to fully understand tsunamigenesis along the Sunda margin.

Lastly, the observed clustering of earthquakes in the shallow forearc mantle is potentially related to the detachment of seamounts at this depth. Again, local earthquake monitoring with a dense station spacing would be a pre-requisite to fully resolve this issue. Detailed hypocenter determination, based on double difference methods, would help to reveal the source of the mantle wedge seismicity.

Appendix A

Appendix of inverted wide-angle traveltimes

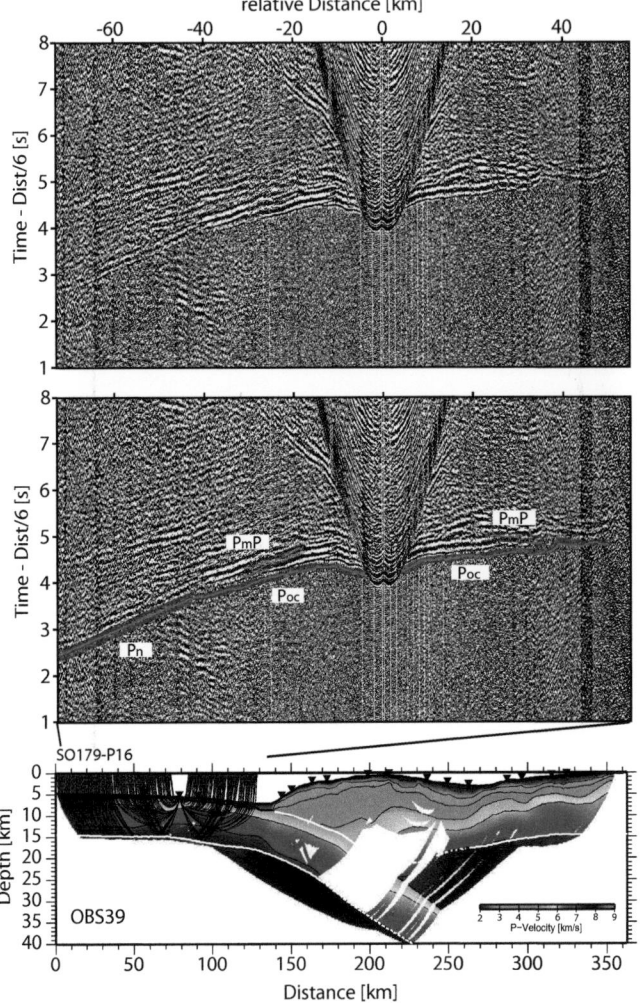

Figure A.1: Inverted model of OBS39

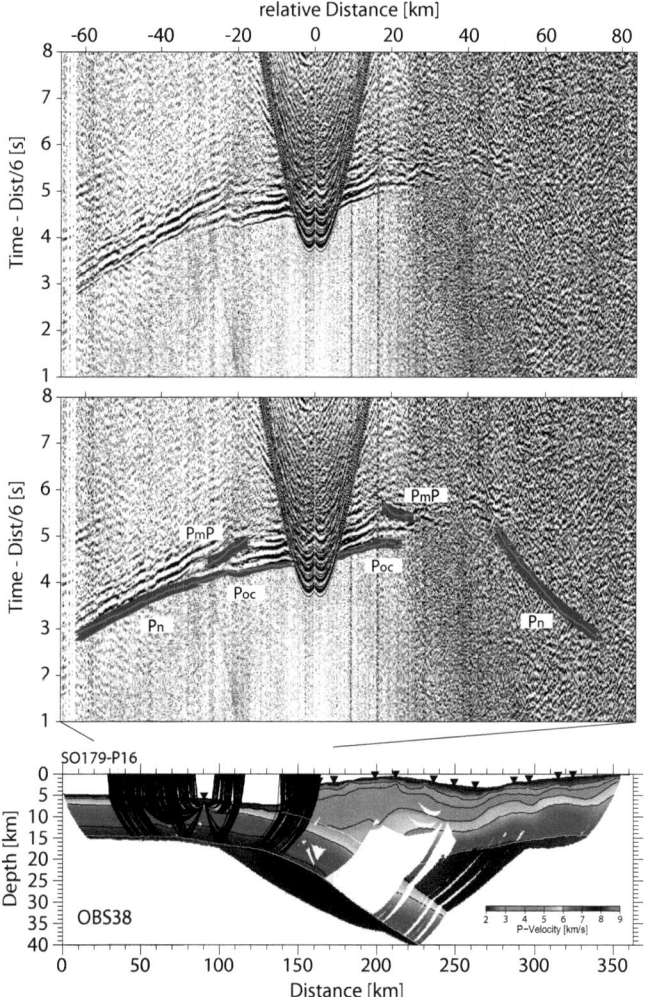

Figure A.2: Inverted model of OBS38

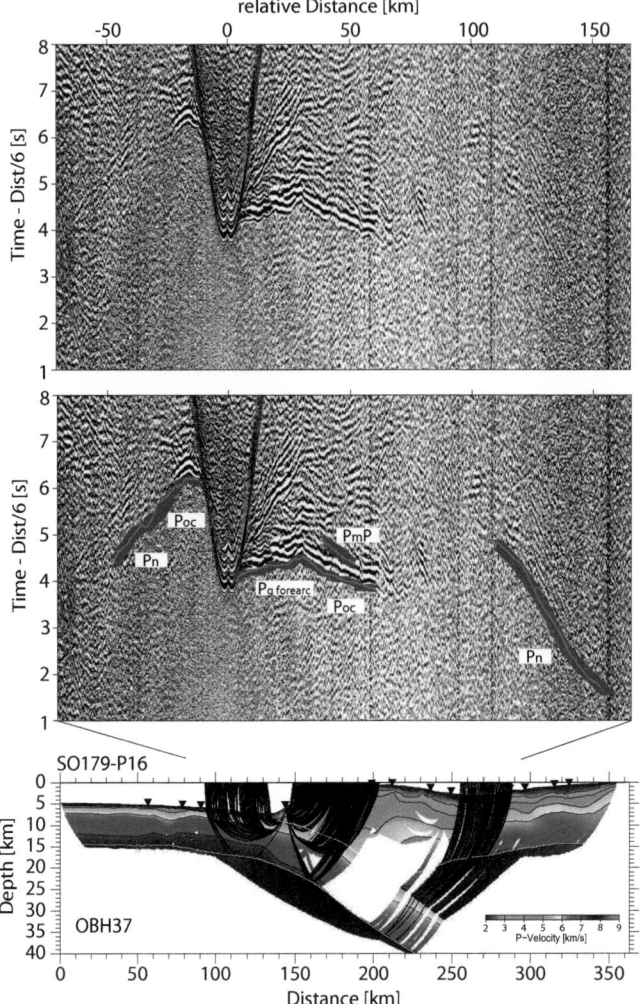

Figure A.3: Inverted model of OBH37

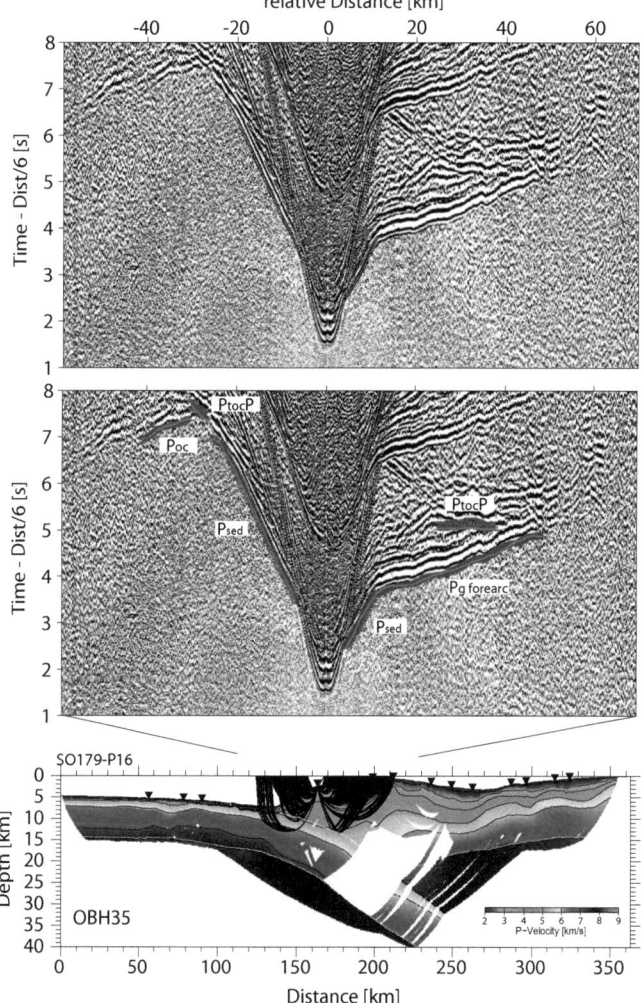

Figure A.4: Inverted model of OBH35

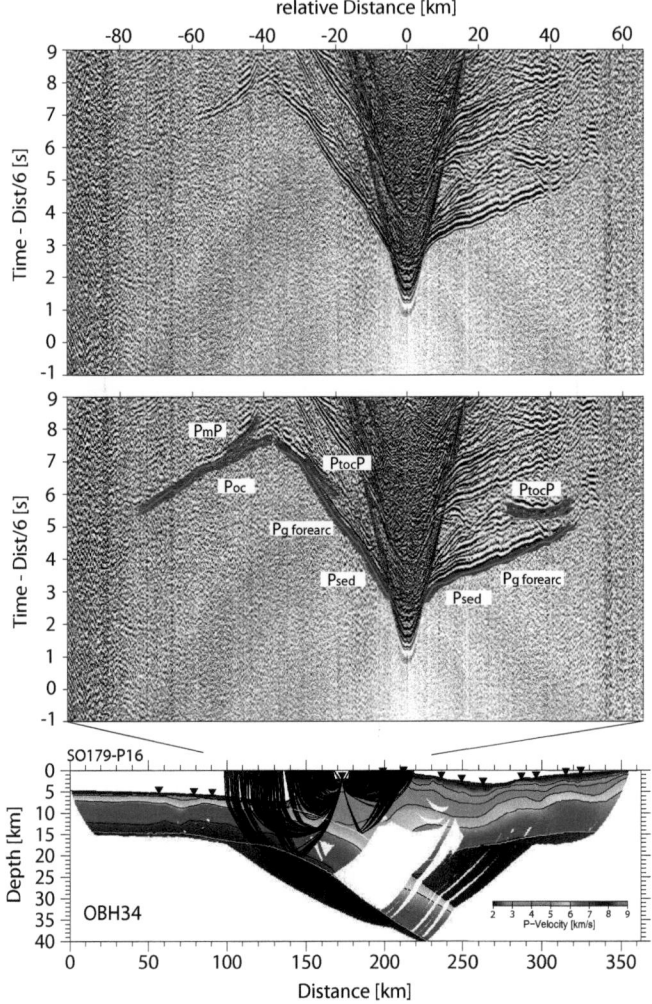

Figure A.5: Inverted model of OBH34

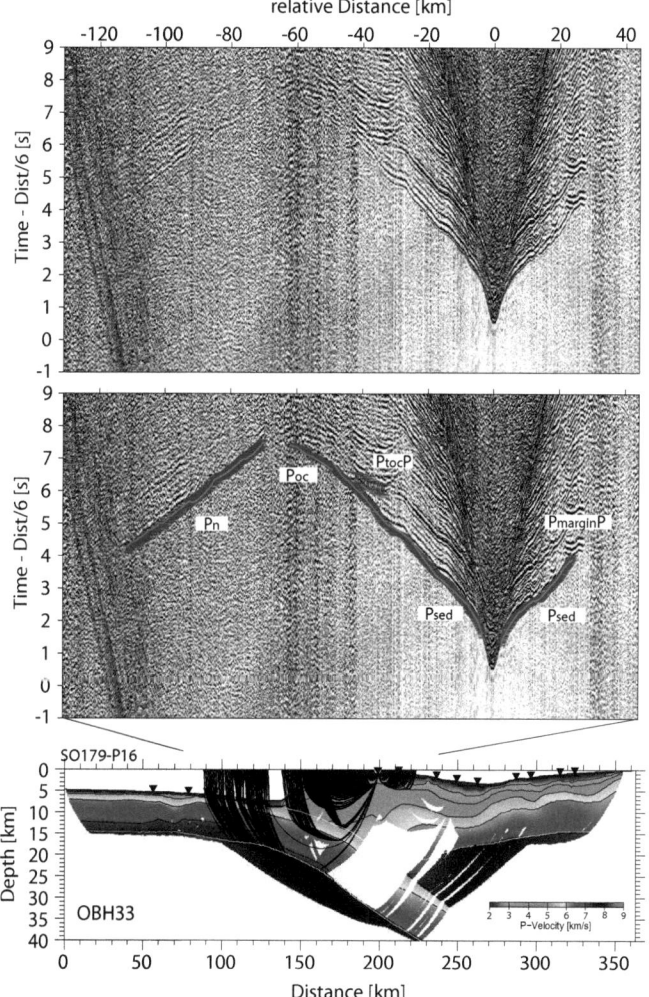

Figure A.6: Inverted model of OBH33

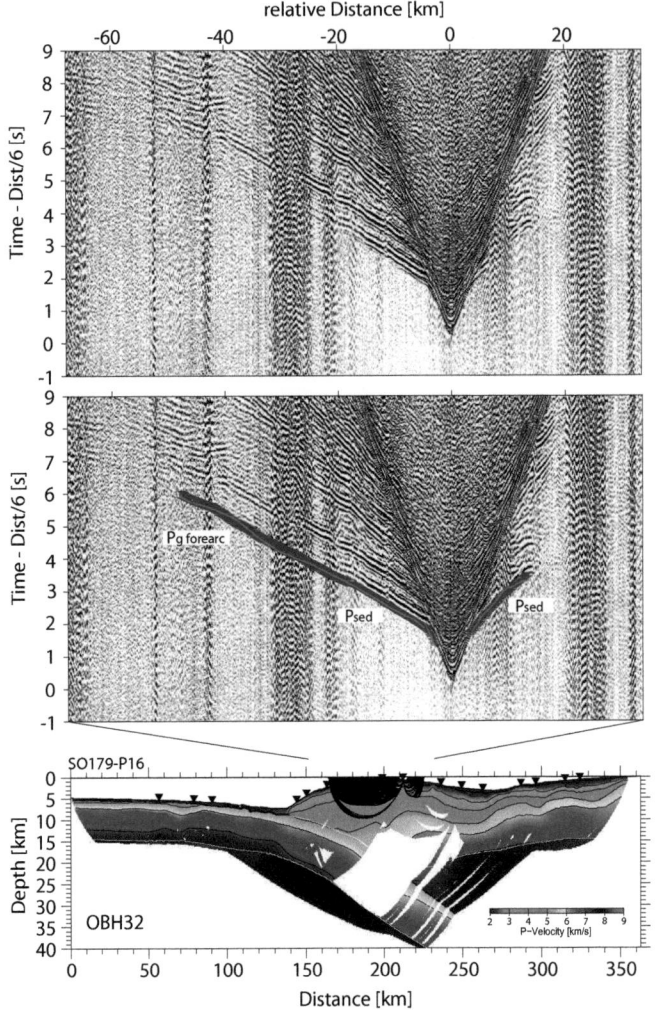

Figure A.7: Inverted model of OBH32

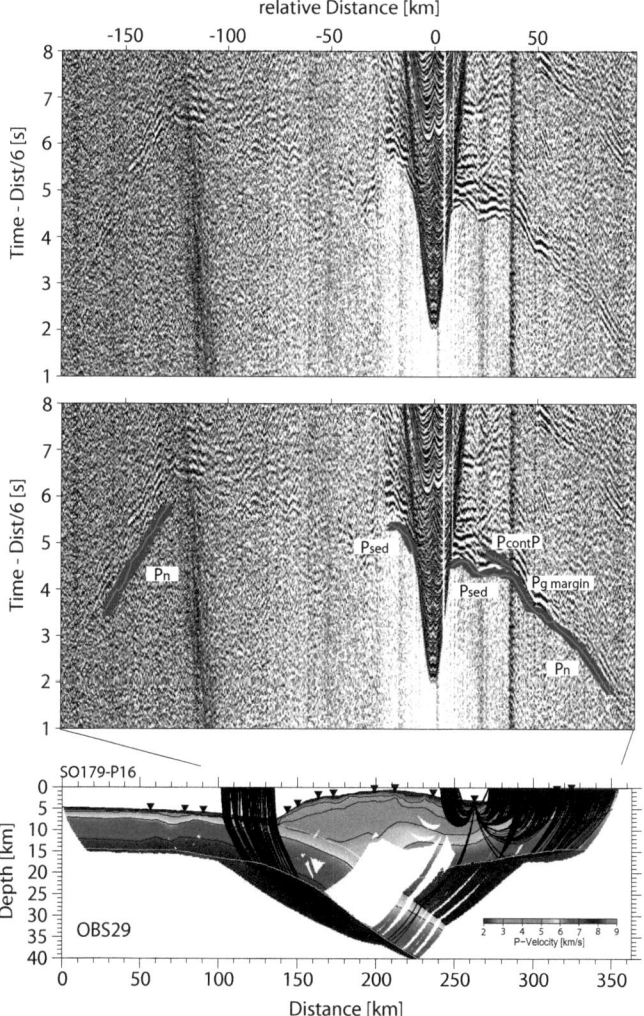

Figure A.8: Inverted model of OBS29

APPENDIX A. APPENDIX OF INVERTED WIDE-ANGLE TRAVELTIMES

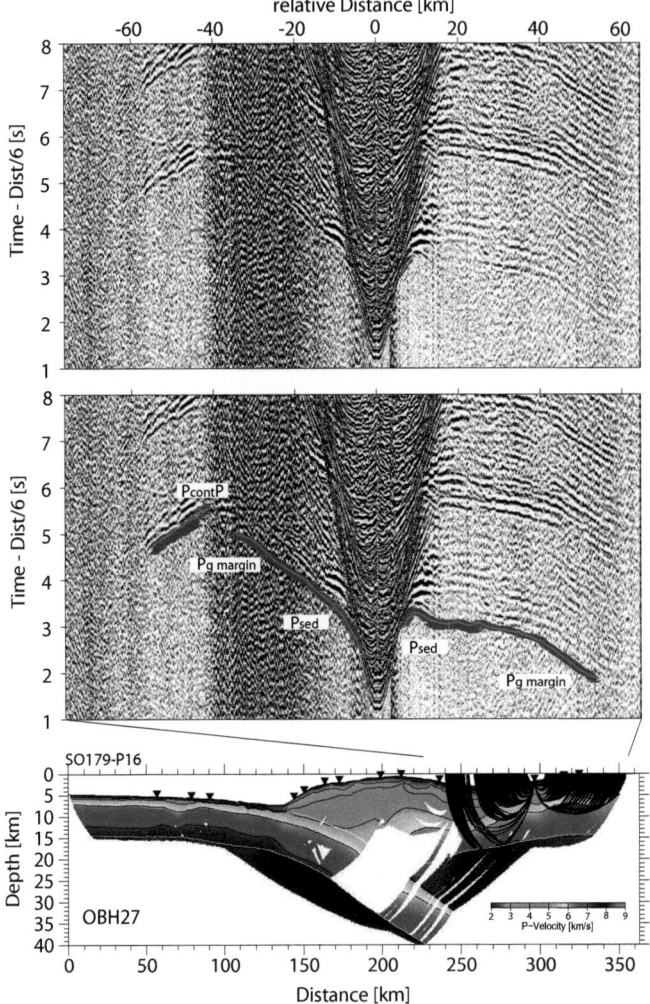

Figure A.9: Inverted model of OBH27

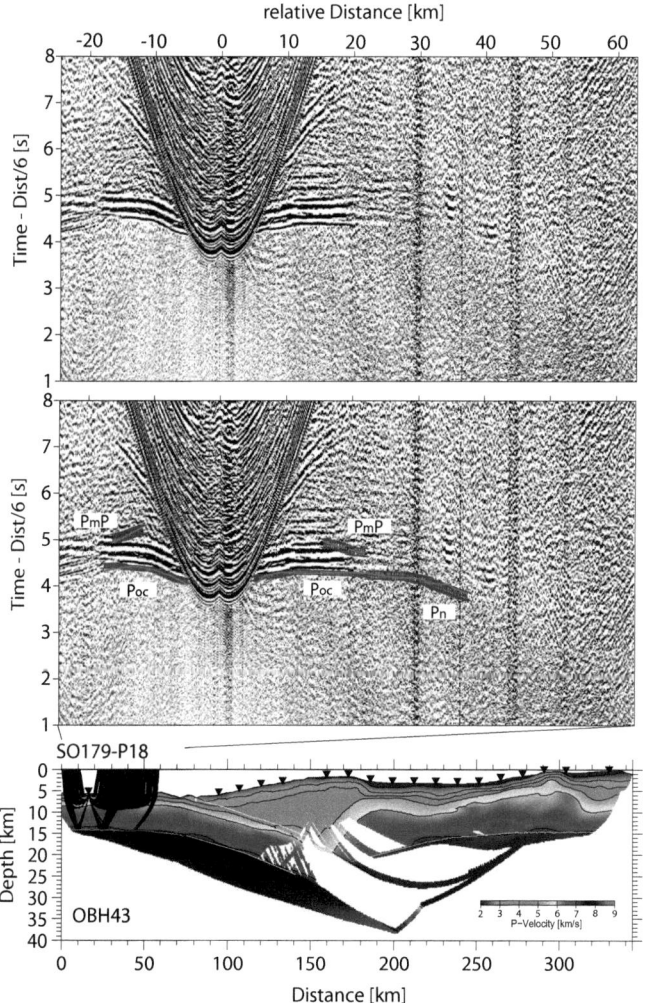

Figure A.10: Inverted model of OBH43

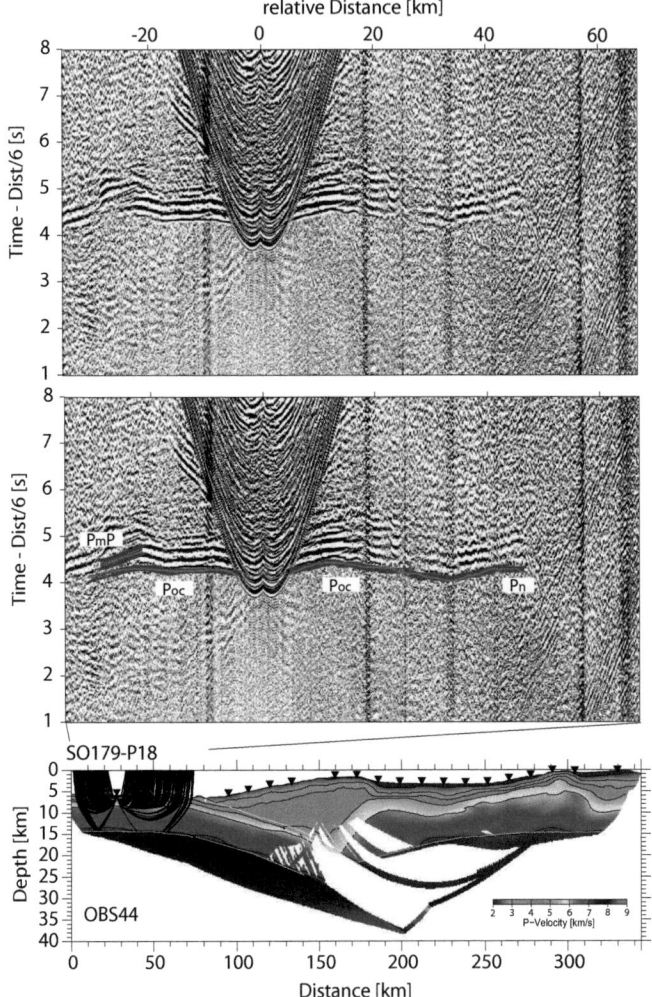

Figure A.11: Inverted model of OBS44

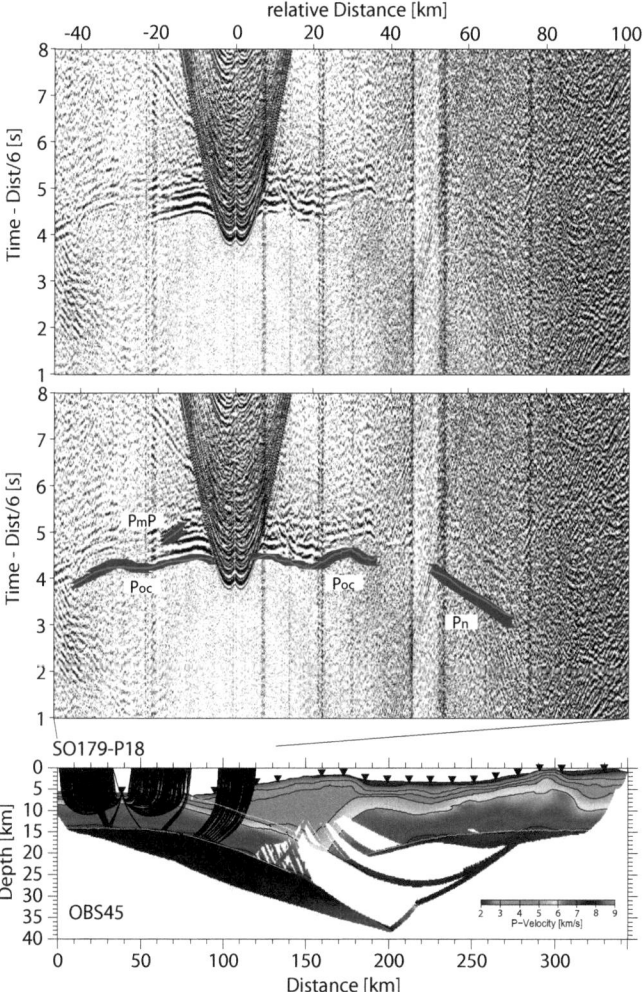

Figure A.12: Inverted model of OBS45

APPENDIX A. APPENDIX OF INVERTED WIDE-ANGLE TRAVELTIMES

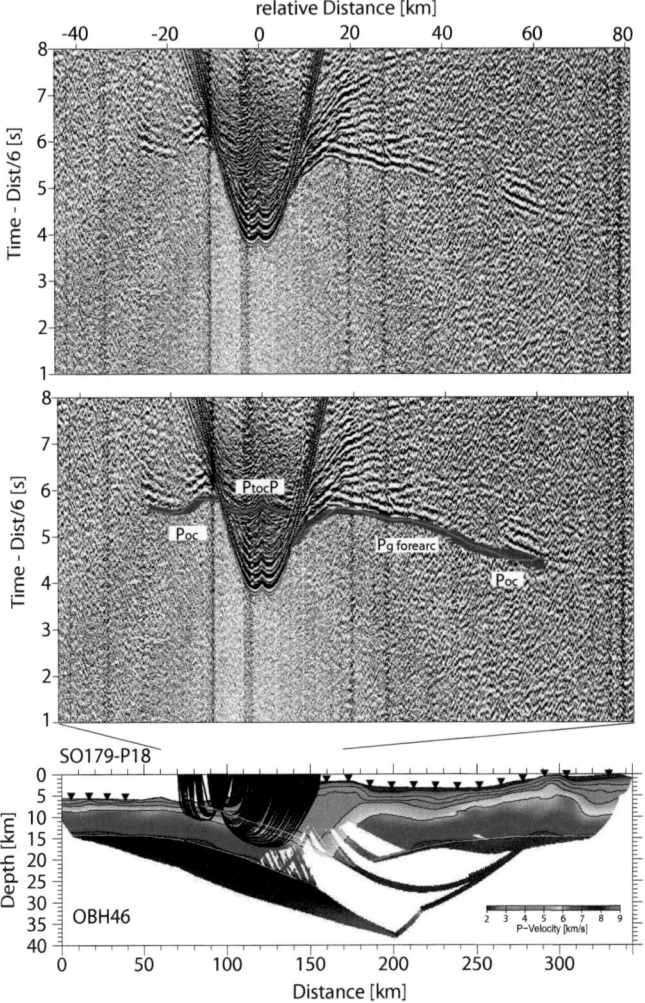

Figure A.13: Inverted model of OBH46

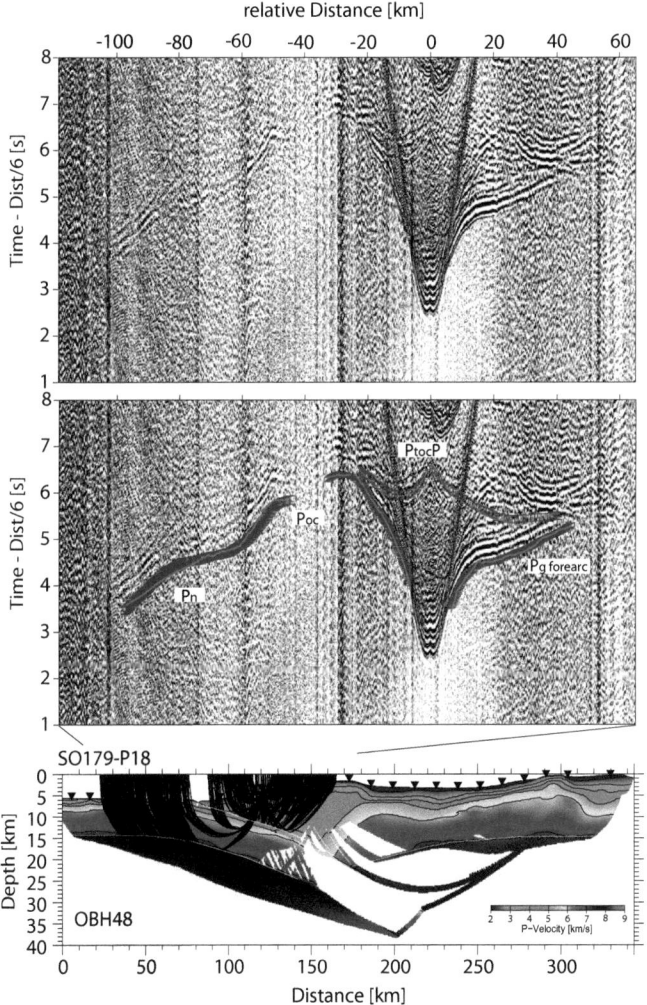

Figure A.14: Inverted model of OBH48

126 APPENDIX A. APPENDIX OF INVERTED WIDE-ANGLE TRAVELTIMES

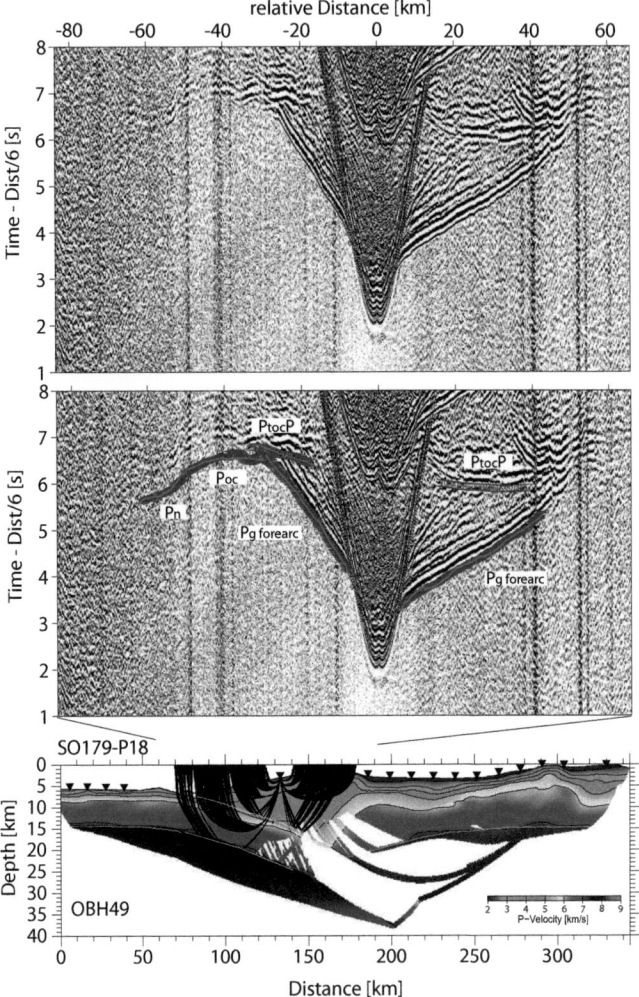

Figure A.15: Inverted model of OBH49

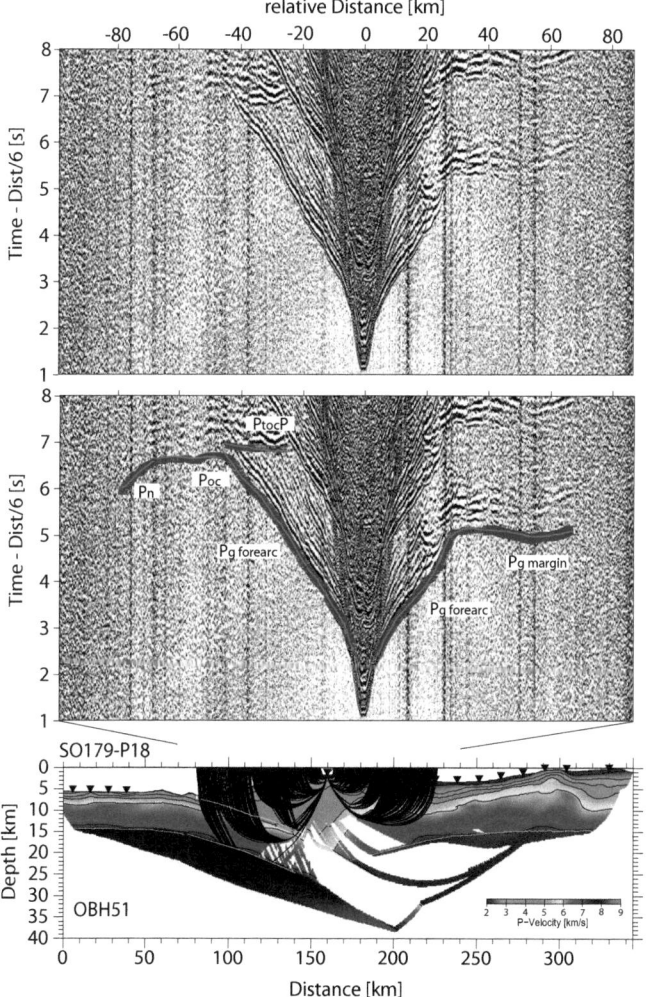

Figure A.16: Inverted model of OBH51

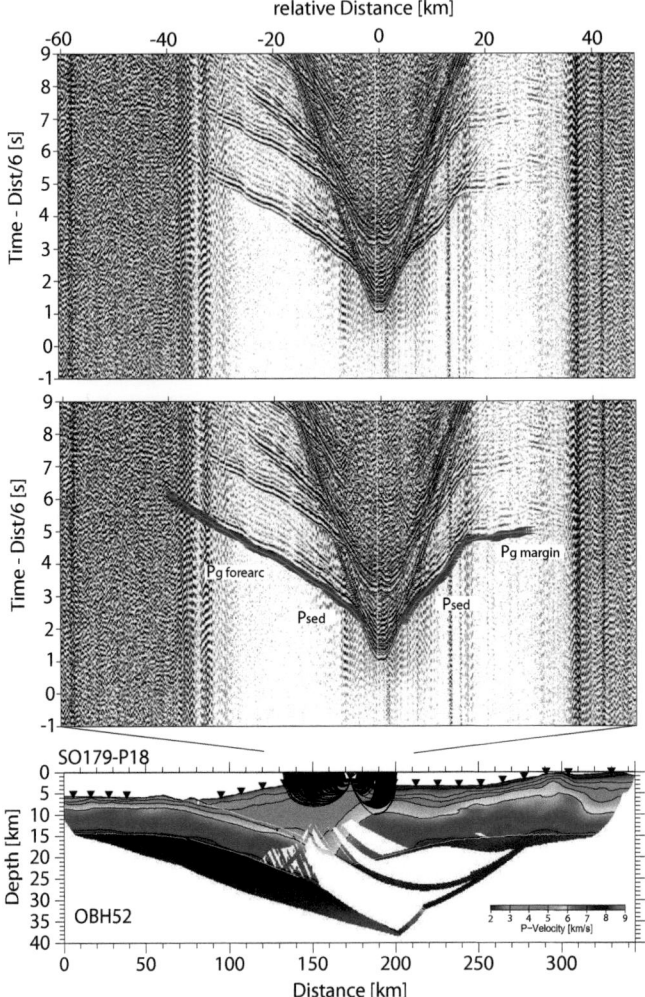

Figure A.17: Inverted model of OBH52

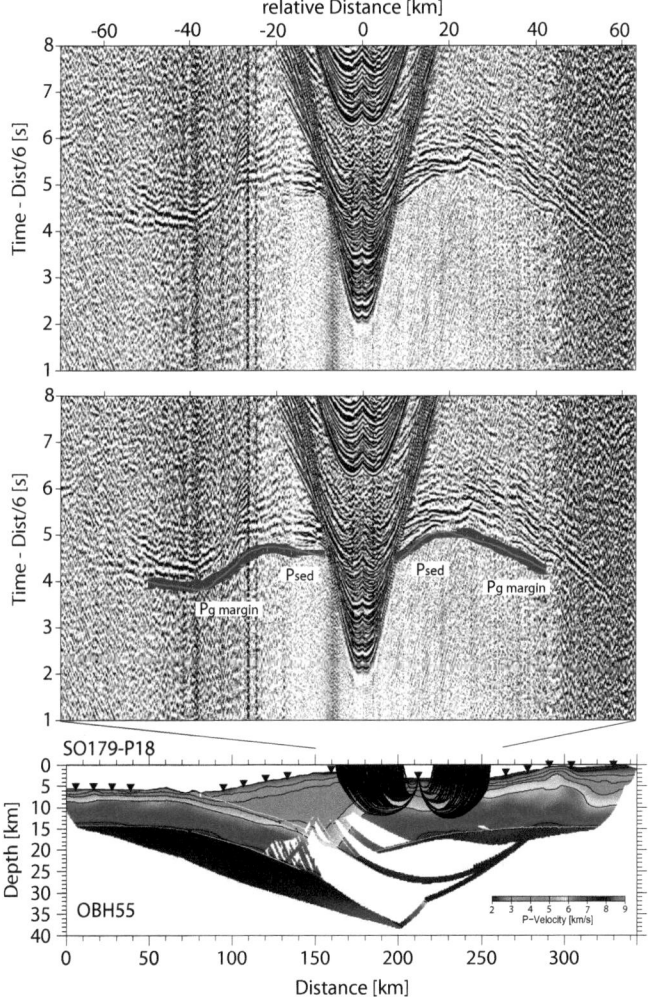

Figure A.18: Inverted model of OBH55

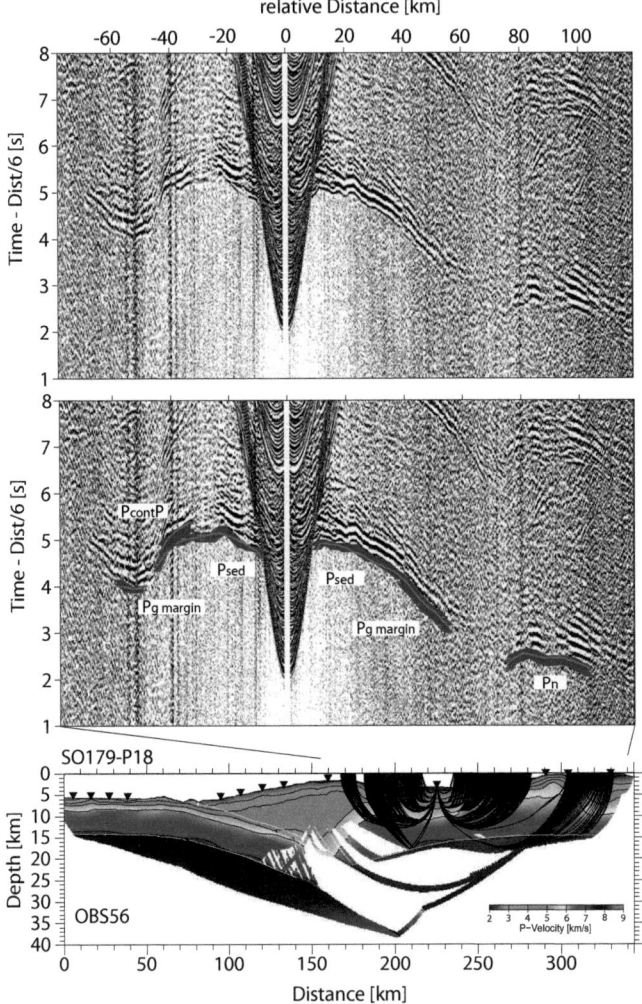

Figure A.19: Inverted model of OBS56

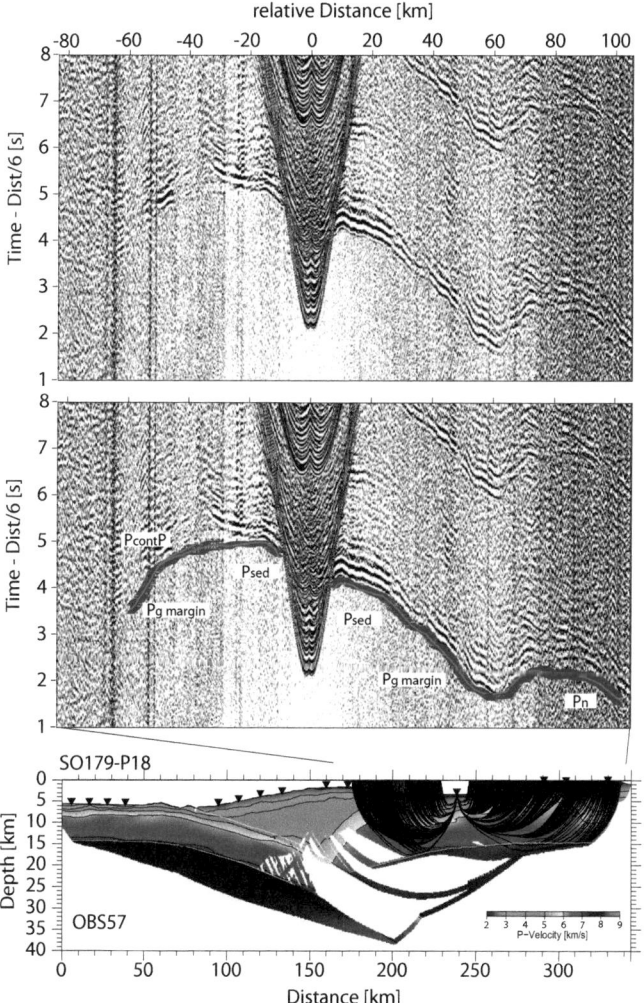

Figure A.20: Inverted model of OBS57

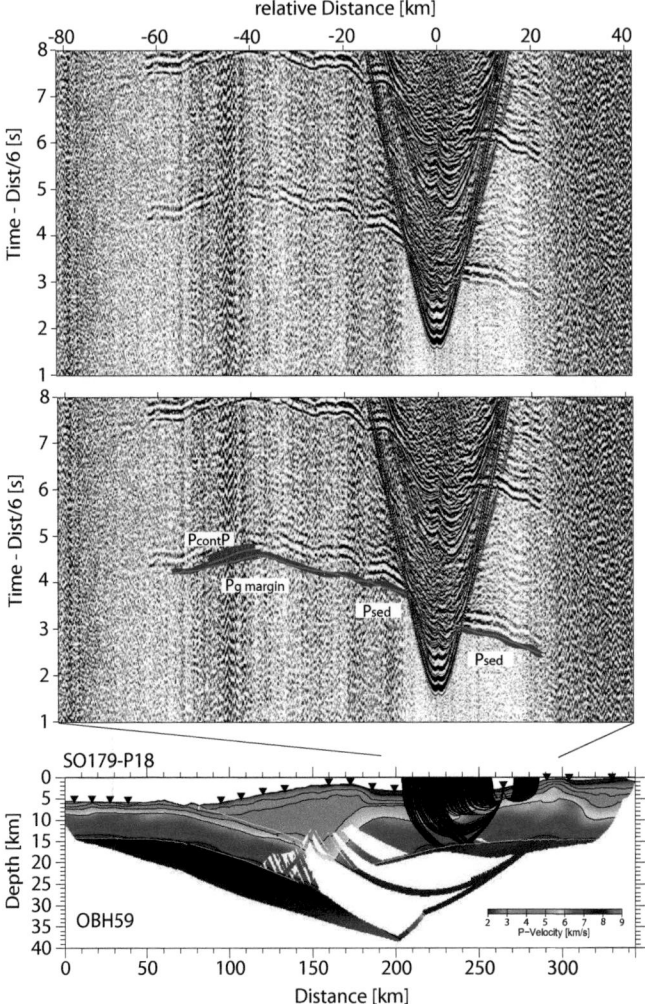

Figure A.21: Inverted model of OBH59

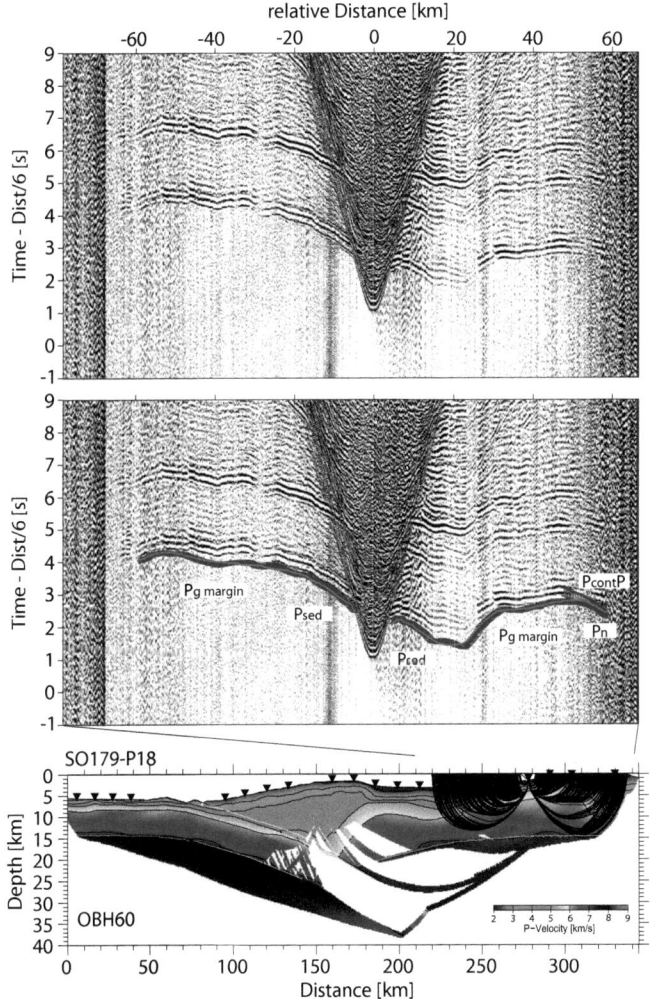

Figure A.22: Inverted model of OBH60

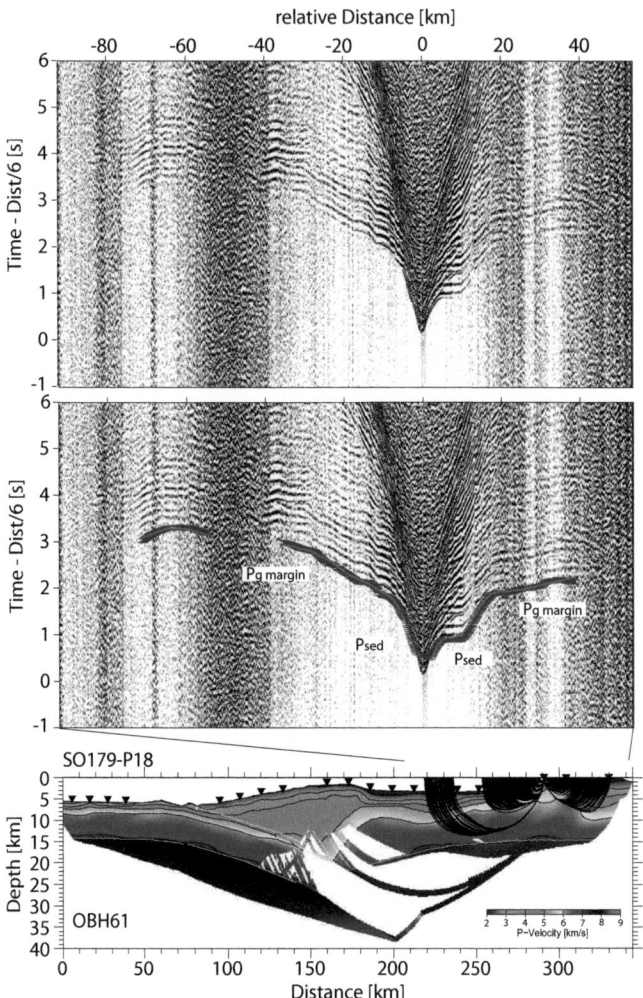

Figure A.23: Inverted model of OBH61

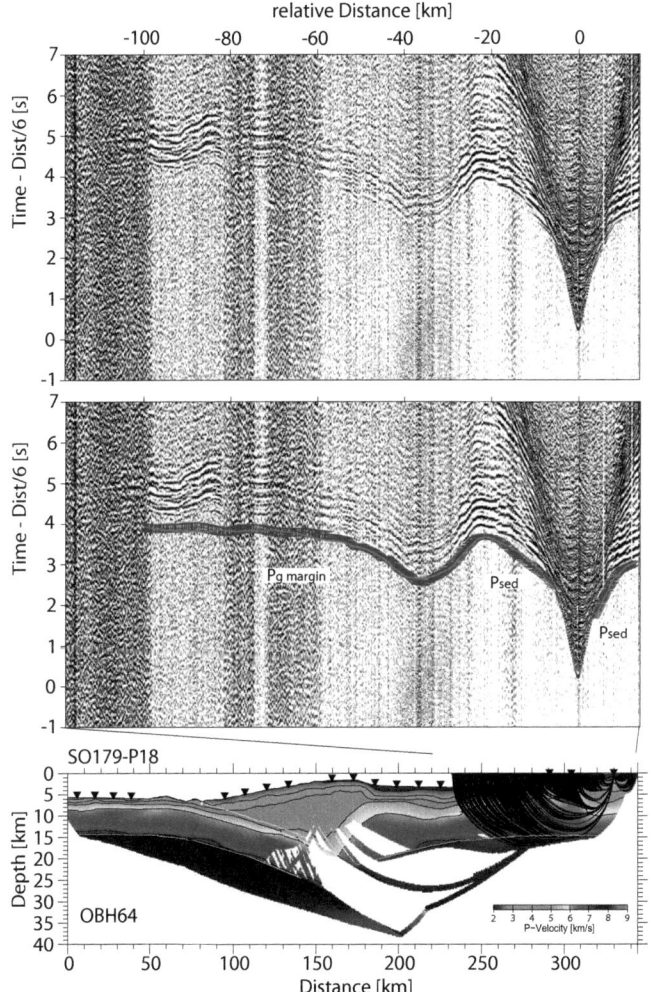

Figure A.24: Inverted model of OBH64

APPENDIX A. APPENDIX OF INVERTED WIDE-ANGLE TRAVELTIMES

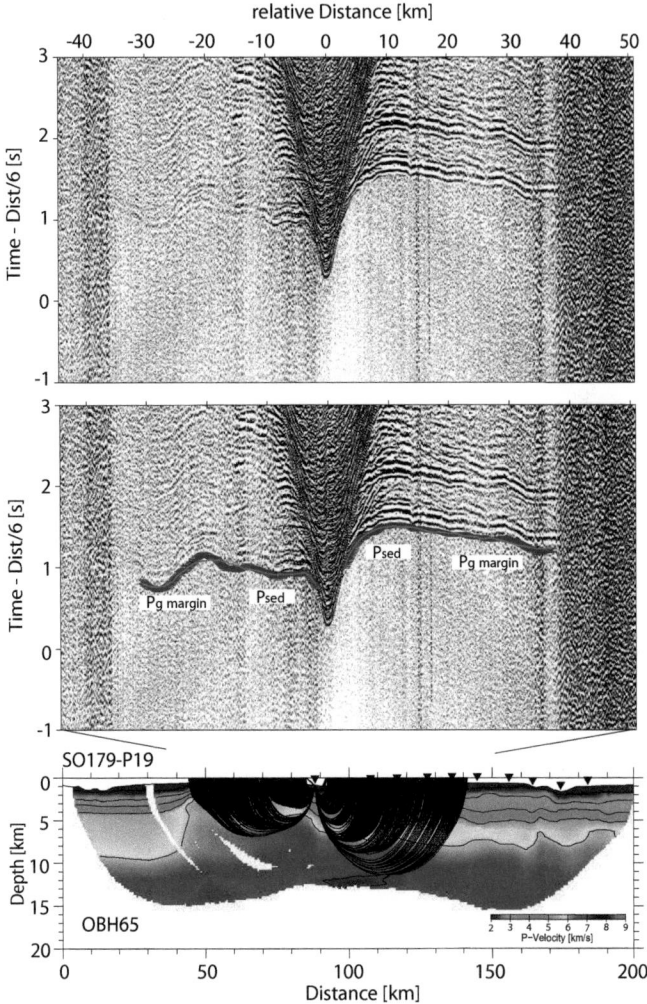

Figure A.25: Inverted model of OBH65

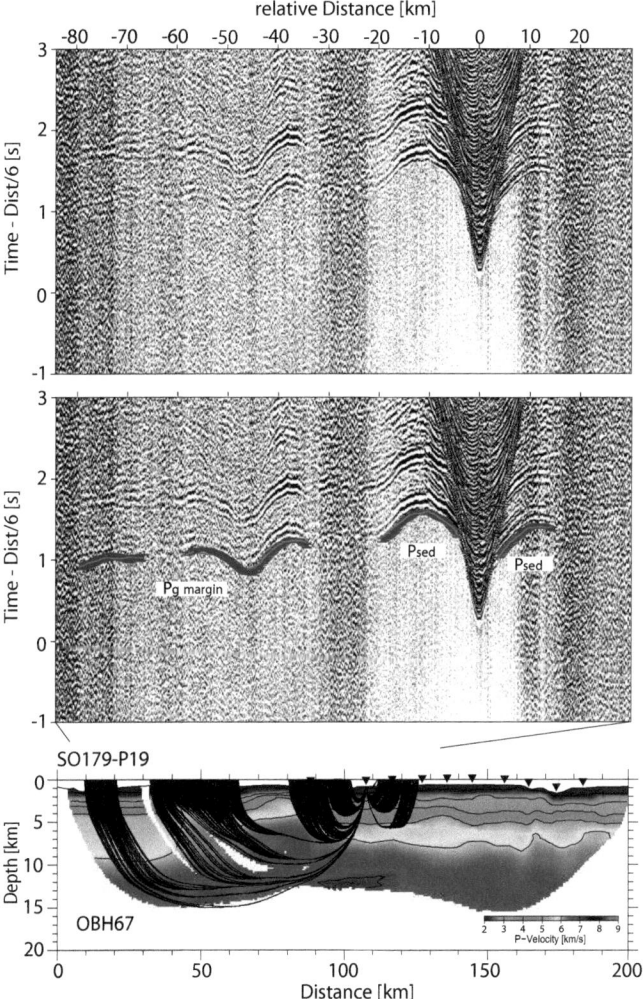

Figure A.26: Inverted model of OBH67

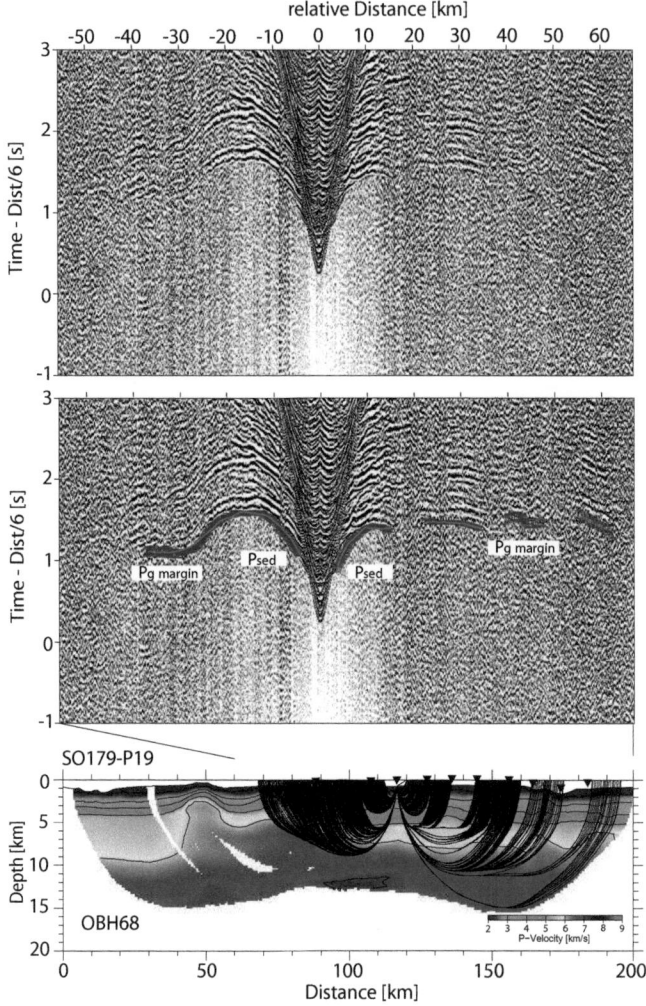

Figure A.27: Inverted model of OBH68

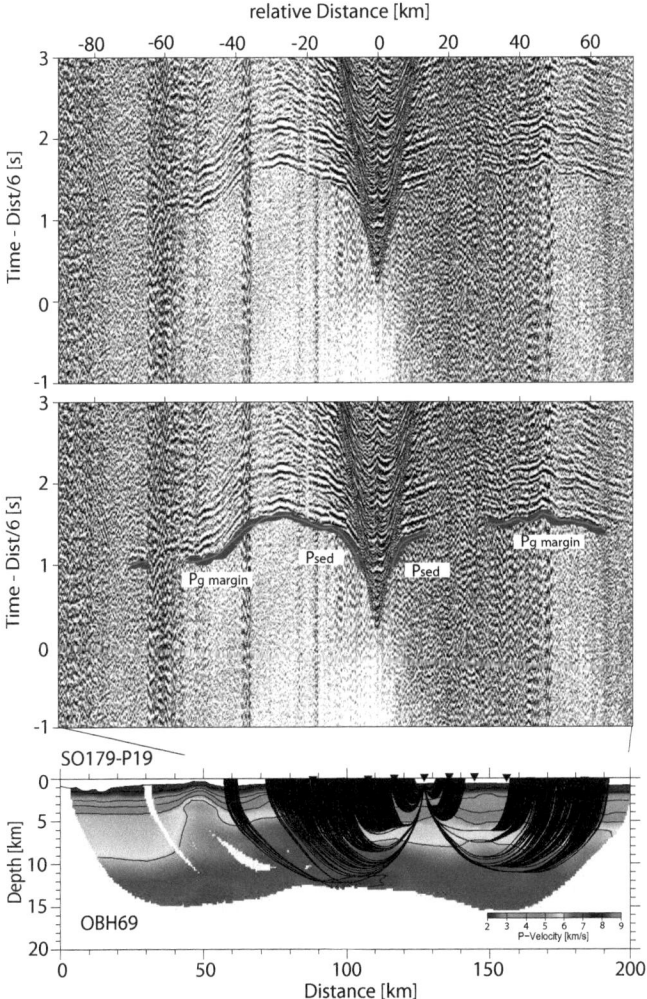

Figure A.28: Inverted model of OBH69

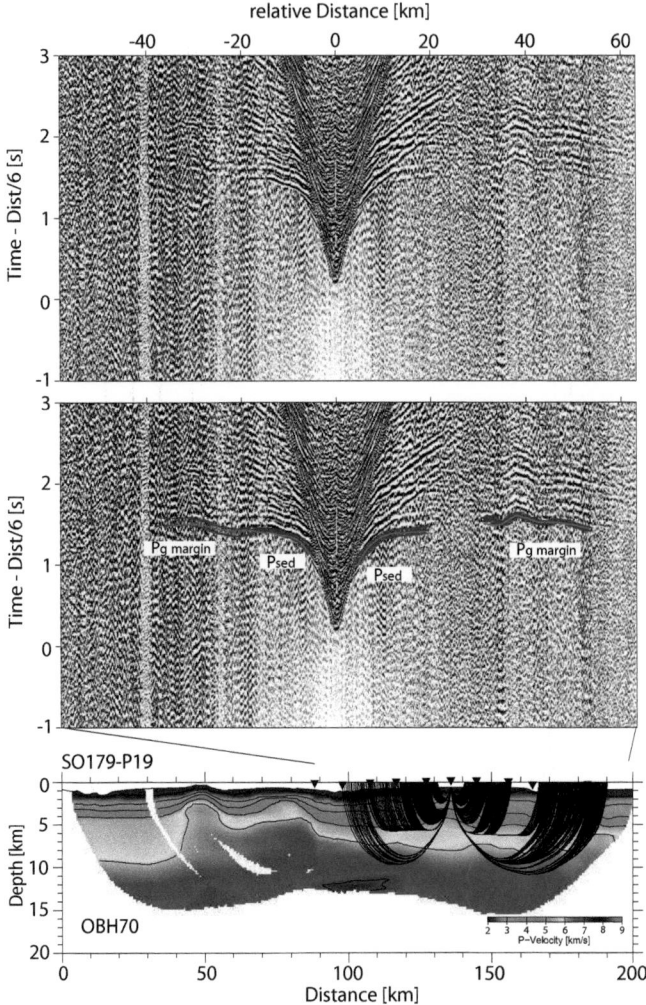

Figure A.29: Inverted model of OBH70

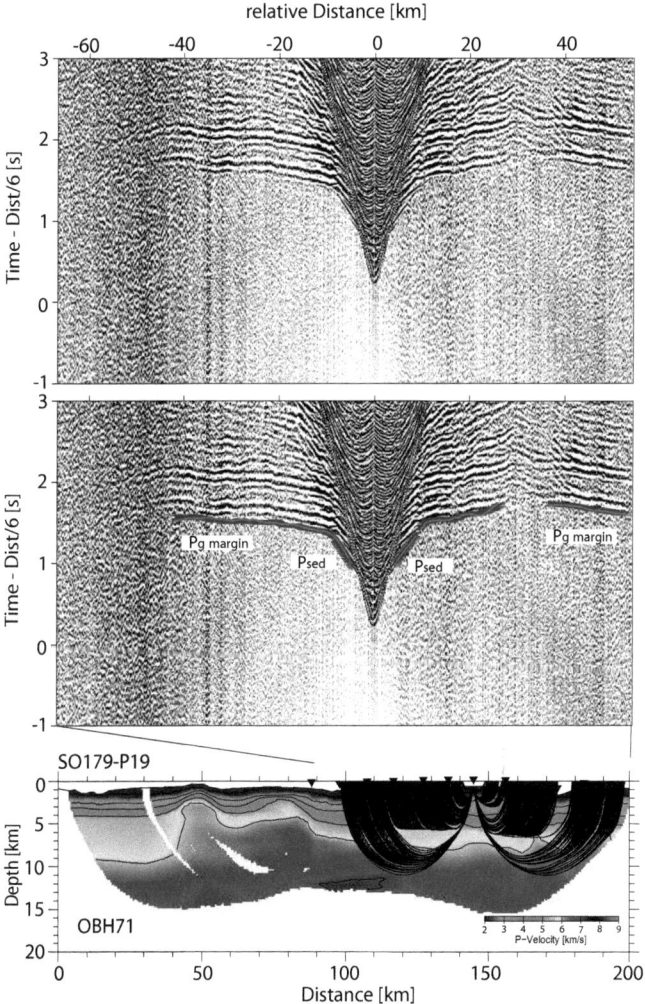

Figure A.30: Inverted model of OBH71

142 APPENDIX A. APPENDIX OF INVERTED WIDE-ANGLE TRAVELTIMES

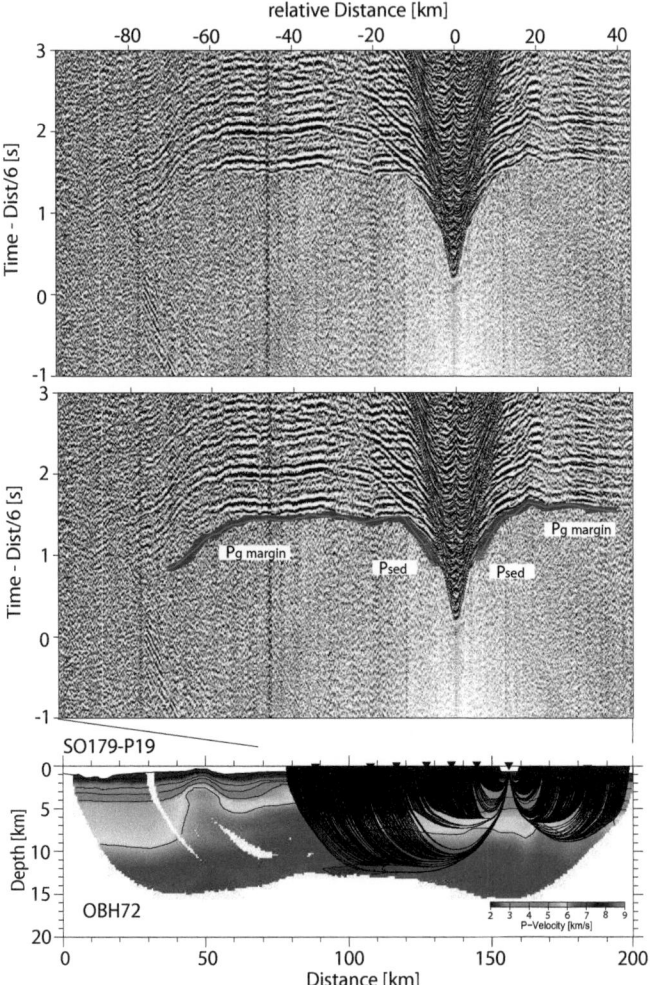

Figure A.31: Inverted model of OBH72

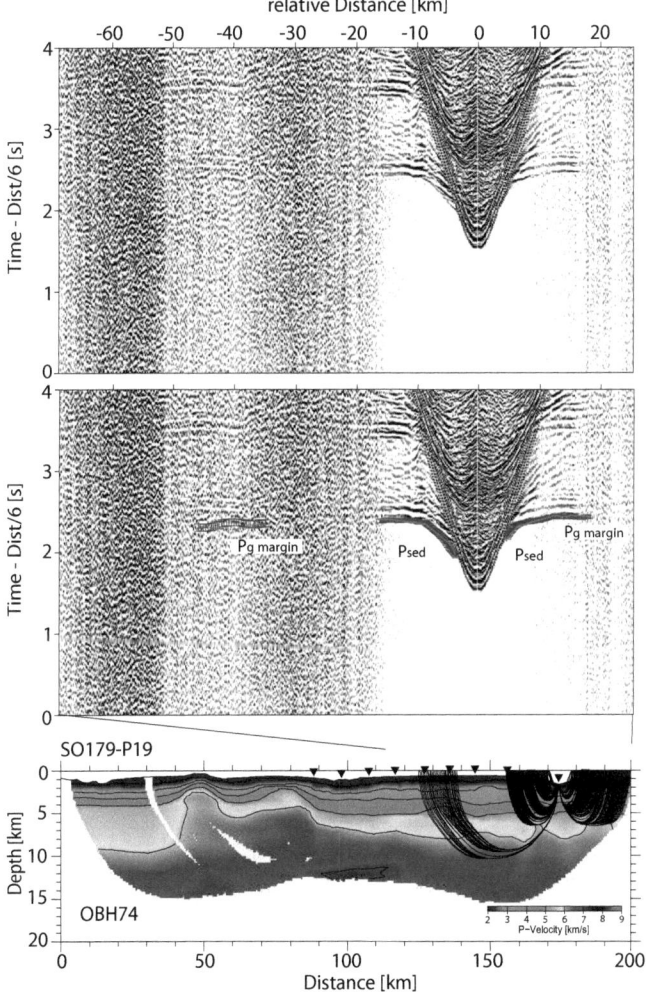

Figure A.32: Inverted model of OBH74

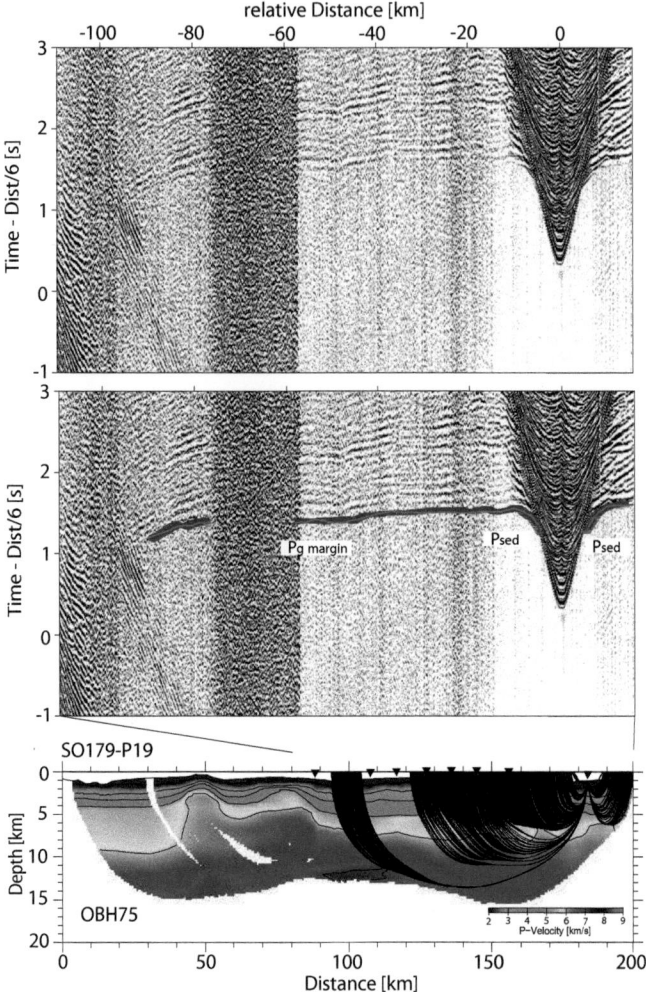

Figure A.33: Inverted model of OBH75

Bibliography

Abercrombie, R., Antolik, M., Felzer, K., and Ekstroem, G. (2001). The 1994 java tsunami earthquake: Slip over a subducting seamount. *Journal of Geophysical Research*, 106(B4):6595–6607.

Ammon, C., Kanamori, H., Lay, T., and Velasco, A. (2006). The 17 july 2006 java tsunami earthquake. *Geophysical Research Letters*, 33:5 PP.

Baba, T., Hori, T., Hirano, S., Cummins, P., Park, J., Kameyama, M., and Kenada, Y. (2001). Deformation of a seamount subducting beneath a accretionary prism: Constraints from numerical simulation. *Geophysical Research Letters*, 28:1827–1830.

Bangs, N., Christeson, G., and Shipley, T. (2003). Structure of the lesser antilles subduction zone backstop and its role in a large accretionary system. *Journal of Geophysical Research*, 108(B7):14 PP.

Bangs, N., Westbrook, G., Ladd, J., and Buhl, P. (1990). Seismic velocities from the barbados ridge complex: Indicators of high pore fluid pressures in an accretionary complex. *Journal of Geophysical Research*, 95:8767–8782.

Barton, P. (1986). The relationship between seismic velocity and density in the continental crust - a useful constraint? *Geophysical Journal of the Royal Astronomical Society*, 87:195–208.

Becel, A., Laigle, M., Diaz, J., Hirn, A., Flueh, E., and Charvis, P. (2010). Searching for conditions of observation of subduction seismogenic zone transients on ocean bottom seismometers deployed at the lesser antilles submerged fore-arc. *Geophysical Research Abstracts*, 12:EGU2010–PREVIEW.

Benaron, N. (1982). A geophysical study of the forearc region south of java, indonesia. Master's thesis, University of San Diego, CA.

Bickel, S. (1990). Velocity-depth ambiguity of reflection traveltimes. *Geophysics*, 55:266–276.

Bilek, S. and Engdahl, E. (2007). Rupture characterization and aftershock relocations for the 1994 and 2006 tsunami earthquakes in the java subduction zone. *Geophysical Research Letters*, 34:5 PP.

BIBLIOGRAPHY

Bishop, T., Bube, K., Cutler, R., Langan, R., Love, P., Resnick, J., Shuey, R., Spindler, D., and Wyld, H. (1985). Tomographic determination of velocity and depth in laterally varying media. *Geophysics*, 50:903–923.

Bollinger, W. and de Ruiter, P. (1976). Geology of the south central java offshore area. *Proceedings of the 45th annual meeting of Indonesian Petroleum Association*, 5:67–81.

Brune, S., Babeyko, A., Ladage, S., and Soboloev, S. (2010). Landslide tsunami hazard in the indonesian sunda arc. *Natural Hazards and Earth Sciences*, 10:589Ű604. submitted.

Brune, S., Ladage, S., Babeyko, A., Mueller, C., Kopp, H., and Soboloev, S. (2009). Submarine landslides at the eastern sunda margin: observations and tsunami impact assessment. *Natural Hazards and Earth Sciences*, 54:547 – 562.

Burbidge, D., Cummins, P., Mleczko, R., and Thio, H. (2008). A probabilistic tsunami hazard assesment for western australia. *Pure and Applied Geophysics*, 165:2059 – 2088.

Byrne, D., Wang, W., and Davis, D. (1993). Mechanical role of backstops in the growth of forearcs. *Tectonics*, 12:123–144.

Carlson, R. and Miller, D. (2003). Mantle wedge water contents estimated from seismic velocities in partially serpentinized peridotites. *Geophysical Research Letters*, 30(5):4 PP.

Carlson, R. and Raskin, G. (1984). Density of the oceanic crust. *Nature*, 311:555–558.

Clift, P. and Vannucchi, P. (2004). Controls on tectonic accretion versus erosion in subduction zones: implications for the origin and recycling of the continental crust. *Review of Geophysics*, 42:31 PP.

Cloos, M. (1992). Thrust-type subduction zone earthquakes and seamount asperities: A physical model for seismic rupture. *Geology*, 82(36):5658–5670.

Contreras-Reyes, E. and Osses, A. (2010). Lithospheric flexure modelling seaward of the chile trench: implications for oceanic plate weakening in the trench outer rise region. *Geophysical Journal International*, 182:97–112.

Curray, J. (1989). The sunda arc: A model for oblique plate convergence. In *Proc. Snellius-II Symp.*, pages 131–140. Neth. Jl. Sea Res.

Curray, J., Shor, G., Russell, J., Raitt, W., and Henry, M. (1977). Seismic refraction and reflection studies of crustal structure of the eastern sunda and western banda arcs. *Journal of Geophysical Research*, 82(17):2479–2489.

Dahlen, F. (1990). Critical taper model of fold-and-thrust belts and accretionary wedges. *Annu. Rev. Earth Planet. Sci.*, 18:55–99.

Das, S. and Watts, A. (2009). *Effect of subducting seafloor topography on the rupture characteristics of great subduction zone earthquakes*. Subduction Zone Geodynamics. Springer Verlag, Berlin-Heidelberg.

Davis, D. (1996). *Accretionary mechanics with properties that vary in space and time*, volume 96 of *Geophysical Monograph*. American Geophysical Union, subduction top to bottom edition.

Davis, D., Suppe, J., and Dahlen, F. (1983). Mechanics of fold-and-thrust belts and accretionary wedges. *Journal of Geophysical Research*, 88:1153–1172.

Davis, E. and Hyndman, R. (1989). Accretion and recent deformation of sediments along the northern cascadia subduction zone. *International Geological Society of America Bulletin*, 101:1465–1480.

Diament, M., Harjono, H., Karta, K., Deplus, C., Dahrin, D., Zen, M., Gerard, M., Lassal, O., Martin, A., and Malod, J. (1992). Mentawai fault zone off sumatra: a new key to the geodynamics of western indonesia. *Geology*, 20:259–262.

Dominguez, J., Malavieille, J., and Lallemand, S. (2000). Deformation of accretionary wedges in response to seamount subduction: insights from sandbox experiments. *Tectonics*, 19(1):182–196.

Dziewonski, A., Ekstroem, G., and Salganik, M. (1995). Centroid-moment tensor solutions for april-june 1994. *Phys. Earth Planet. Inter.*, 88:69–78.

Fischer, R. and Lees, J. (1993). Shortest path ray tracing with sparse graphs. *Geophysics*, 58:987–996.

Font, Y. and Lallemand, S. (2009). Subducting oceanic high causes compressional faulting in southernmost ryukyu forearc as revealed by hypocentral determinations of earthquakes and reflection/refraction seismic data. *Tectonophysics*, 466:255–267.

Fruehn, J., von Huehne, and Fisher, M. (1999). Accretion in the wake of terrane collision: the neogene accretionary wedge off kenai peninsula, alaska. *Tectonics*, 18(2):263–277.

Fuller, C., Willet, S., Fisher, D., and Lu, C. (2006). A thermomechanical wedge model of taiwan constrained by fission-track thermochronometry. *Tectonophysics*, 425:1–24.

Ganie, B., Syafuddin, Superman, A., and Honza, E. (1987). Geomorphological features in the eastern sunda trench. *CCOP Technical Bulletin*, 19:7–12.

Gerya, T., Fossati, D., Cantieni, C., and Seward, D. (2009). Dynamic effects of aseismic ridge subduction: numerical modeling. *European Journal of Mineralogy*, 21:649–661.

Ghose, R., Yoshioka, S., and Oike, K. (1990). Three-dimensional numerical simulation of the subduction dynamics in the sunda arc region, southeast asia. *Tectonophysics*, 181:223–255.

Graindorge, D., Calahorrano, A., Charvis, P., Collot, J.-Y., and Bethoux, N. (2004). Deep structures of the ecuador convergent margin and the carnegie ridge, possible consequence on great earthquake recurrence interval. *Geophysical Research Letters*, 31:5 PP.

Grevemeyer, I. and Tiwari, V. (2006). Overriding plate controls spatial distribution of megathrust earthquakes in the sunda-andaman subduction zone. *Earth and Planetary Science Letters*, 251:199–208.

Hamilton, W. (1979). Tectonics of the indonesian region. *U.S. Geological Survey Prof. Pap.*, 1078:308–335.

Hamilton, W. (1988). Plate tectonics and island arcs. *Geol. Soc. Am. Bull.*, 100:1053–1527.

Hino, R., Shiobara, S., Shimamura, H., Sato, S., Kanazawa, T., Kasahara, J., and Hasegawa, A. (2000). Aftershock distribution of the 1994 sanriku-oki earthquake (mw 7.7) revealed by ocean bottom seismographic observation. *Journal of Geophysical Research*, 105(B9):21697 – 21710.

Huchon, P. and Le Pichon, X. (1984). Sunda strait and central sumatra fault. *Geology*, 12:668–672.

Hughes, R. and Pilatasing, L. (2002). Cretaceous and tertiary terrane accretion in the cordillera occidental of the andes of ecuador. *Tectonophysics*, 345(1-4):29–48.

Hyndman, R., Yamano, M., and Oleskevich, D. (1997). *The seismogenic zone of subduction thrust faults*, volume 6 of *The island arc*. Island Arc.

Hyndman, R., Yorath, C., Clowes, R., and Davis, E. (1990). The northern cascadia subduction zone at vancouver island: seismic structure and tectonic history. *Can. J. Earth Sci.*, 27(2):313–329.

Jiao, W., Silver, P., Fei, Y., and Prewitt, C. (2000). Do intermediate- and deep-focus earthquakes occur on preexisting waek zones? an examination of the tonga subduction zone. *Journal of Geophysical Research*, 105(28):125–138.

Kanamori, H. (1972). Mechanism of tsunami earthquakes. *Phys. Earth Planet. Inter.*, 6:346–359.

Karig, D., Suparka, S., Moore, G., and Hehunassa, P. (1979). Strucutre and cenozoic evolution of the sunda arc in the central sumatra region. *American Association of Petr. Geol. Mem.*, 29:223–237.

Kirby, S., Engdahl, E., and Denlinger, R. (1996). *Intermediate-depth intraslab earthquakes and arc volcanism as physical expressions of crustal and uppermost mantle metamorphism in subducted slabs*, volume 96 of *Geophysical Monograph Series*. American Geophysical Union, subduction top to bottom edition.

Klingelhoefer, F., Gutscher, M., Ladage, S., Dessa, J., Graindorge, D., Franke, D., Andre, C., Permana, H., Yudistira, T., and Chauhan, A. (2010). Limits of the seismogenic zone in the epicentral region of the 26 december 2004 great sumatra-andaman earthquake: Results from seismic refraction and wide-angle reflection surveys and thermal modeling. *Journal of Geophysical Research*, 115:23 PP.

Kopp, H. and Flueh, E. (2004). *FS Sonne Cruise Report SO176/179 MERAMEX I/II*, volume 1 of *IFM-GEOMAR Report*. Leibnitz-Institut für Meereswissenschaften and der Christian-Albrechts-Universität zu Kiel.

Kopp, H., Flueh, E., Klaeschen, D., Bialas, J., and Reichert, C. (2001). Crustal structure of the central sunda margin at the onset of oblique subduction. *Geophysical journal International*, 147(2):449–474. 10.1046/j.0956-540x.2001.01547.x.

Kopp, H., Flueh, E., Papenberg, C., Klaeschen, D., and the SPOC-Scientists (2004). Seismic investigations of the o'higgins seamount group and juan fernandez ridge. *Tectonics*, 23(23):21 PP.

Kopp, H., Flueh, E., Petersen, C., Weinrebe, W., Wittwer, A., and the MERAMEX Scientists (2006). The java margin revisited: Evidence for subduction erosion off java. *Earth and Planetary Science Letters*, 242:130–142.

Kopp, H., Hindle, D., Oncken, O., Reichert, C., and Scholl, D. (2009). Anatomy of the western java plate interface from depth-migrated seismic images. *Earth and Planetary Science Letters*, 288:399–407.

Kopp, H., Klaeschen, D., Flueh, E., and Bialas, J. (2002). Crustal structure of the java margin from seismic wide-angle and multichannel reflection data. *Journal of Geophysical Research*, 107(B2):2034. 10.1029/2000JB000095.

Kopp, H. and Kukowski, N. (2003). Backstop geometry and accretionary mechanics of the sunda margin. *Tectonics*, 22(6):16 PP. 10.1029/2002TC001420.

Koppers, A. and Watts, A. (2010). Intraplate seamounts as a window into deep earth processes. *Oceanography*, 23(1):42–57.

Korenaga, J., Holbrook, W., Kent, G., Kelemen, P., Detrick, R., Larsen, H., Hopper, J., and Dahl-Jensen, T. (2000). Crustal structure of the southeast greenland margin from joint refraction and reflection seismic tomography. *Journal of Geophysical Research*, 105(21):591–614.

Koulakov, I., Bohm, M., Asch, G., Luehr, B., Manzanares, A., Brotopuspito, K., Fauzi, P., Purbawinata, M., Puspito, N., Ratdomopurbo, A., Kopp, H., Rabbel, W., and Shevkunova, E. (2007). P and s velocity structure of the crust and the upper mantle beneath central java from local tomography inversion. *Journal of Geophysical Research*, 112(B08310):19 PP. 10.1029/2006JB004712.

Lallemand, E. (1994). Coulomb theory applied to accretionary and nonaccretionary wedges: Possible causes for tectonic erosion and/or frontal accretion. *Journal of Geophysical Research*, 99(6):12033–12055.

Le Pichon, X. and Henry, P. (1992). Erosion and accretion along subduction zones: A model of evolution. *Proc. K. Ned. Akad. Wet.*, 95:297–310.

Litchfield, N., Ellis, S., Berryman, K., and Nicol, A. (2007). Insights into subduction-related uplift along the hikurangi margin, new zealand, using numerical modeling. *Journal of Geophysical Research*, 112:17 PP.

Lohrmann, J., Kukowski, N., Adam, J., and Oncken, O. (2003). The impact of analogue material properties on the geometry, kinematics and dynamics of convergent sand wedges. *Journal of Structural Geology*, 25(10):1691–1711.

Ludwig, W., Nafe, J., and Drake, C. (1970). *Seismic refraction*, volume 4 of *The sea*. Wiley-Interscience.

Luetgert, J. (1992). Macray-interactive two-dimensional seismic raytracing for the macintosh. *U.S. Geol. Surv. Open File Report*, 92-356:48 PP.

Malod, J. and Kemal, B. (1996). *The Sumatra margin: oblique subduction and lateral displacement of the accretionary prism*, volume 106 of *Tectonic evolution of Southeast Asia*. Geological Society Special Publication.

Masson, D., Parson, L., Milsom, J., Nichols, G., Sikumbang, N., Dwiyanto, B., and Kallagher, H. (1990). Subduction of seamounts at the java trench: a view with long-range sidescan sonar. *Tectonophysics*, 185:51–65.

Menke, W. (1989). Geophysical data analysis: Discrete inverse theory. *International Geophysics Series*, 45:285p.

Mochizuki, K., Yamada, Y., Shinohara, M., Yamanaka, Y., and Kanazawa, T. (2008). Weak interplate coupling by seamounts and repeating m 7 earthquakes. *Science*, 321:1194 – 1197.

Moore, G., Curray, J., and Moore, D. (1980). Variations in in geologic structure along the sunda forearc, northeastern indian ocean. *Geophysical Monographs*, 23:145–160.

Moser, T. (1991). Shortest path calculation of seismic rays. *Geophysics*, 56:59–67.

Moser, T., Van Eck, T., and Nolet, G. (1992). Hypocenter determination in strongly heterogeneous earth models using the shortest path method. *Journal of Geophysical Research*, 97:6563Ű6572.

Mueller, C., Kopp, H., Djajadihardja, Y., Barckhausen, U., Ehrhardt, A., Engels, M., Flueh, E., Gaedicke, C., Keppler, H., Lutz, R., Lueschen, E., Neben, S., Seeber, L., and Dzulkarnaen, D. (2008). From subduction to collision: the sunda-banda arc transition. *EOS*, 89(6):49.

Nakajima, J., Matszawa, T., Hasegawa, A., and Zhao, D. (2001). Three dimensional strucutre of vp, vs and vp/vs beneath northeastern japan: Implications for arc magmatism and fluids. *Journal of Geophysical Research*, 106:21,843–21,858.

Newcomb, K. and McCann, W. (1987). Seismic history and seismotectonics of the sunda arc. *Journal of Geophysical Research*, 92(B1):421–439.

O'Hanley, D. (1996). *Serpentines: Recorders of tectonic and petrological history*. Oxford Monographs on Geology and Geophysics. Oxford University Press, New York.

Oleskevich, D. A., Hyndman, R., and Wang, K. (1999). The updip and downdip limits to great subduction earthquakes: Thermal and structural models of cascadia, south alaska, sw japan, and chile. *Journal of Geophysical Research*, 104(7):14965 – 14991.

Paige, C. and Saunders, M. (1982). An algorithm for sparse linear equations and sparse least squares. *Trans. Math. Software*, 8:43–71.

Peacock, S. (1990). Fluid processes in subduction zones. *Science*, 248:329–337.

Peacock, S. (2001). Are the lower planes of double seismic zones caused by serpentine dehydration in subducting oceanic mantle. *Geology*, 29:299–302.

Planert, L., Kopp, H., Lueschen, E., Mueller, C., Flueh, E., Shulgin, A., Djajadihardja, Y., and Krabbenhoeft, A. (2010). Lower plate structure and upper plate deformational segmentation at the sunda-banda arc transition, indonesia. *Journal of Geophysical Research*, 115:25 PP. submitted.

Polet, J. and Kanamori, H. (2000). Shallow subduction zone earthquakes and their tsunamigenic potential. *Geophysical Journal International*, 142:684–702.

Ranero, C., Morgan, J., McIntosh, K., and Reichert, C. (2003). Bending related faulting and mantle serpentization at the middle america trench. *Nature*, 425:367–373.

Reinen, L., Weeks, J., and Tullis, T. (1994). The frictional behavior of lizardite and antigorite serpentines: Experiments, constitutive models, and implications for natural faults. *Pure and Applied Geophysics*, 143:317 – 358.

Robinson, D., Das, S., and Watts, A. (2006). Earthquakes stalled by a subducting fracture zone. *Science*, 312(5777):1203–1205.

Ryan, H. and Scholl, D. (1989). The evolution of forearc structures along an oblique convergent margin, central aleutian arc. *Tectonics*, 8(3):497–516.

Sallares, V. and Ranero, C. (2005). Structure and tectonics of the erosional convergent margin off antofagasta, north chile. *Journal of Geophysical Research*, 110:19 PP.

Schlueter, H., Gaedicke, C., Roeser, H., Schreckenberger, B., Meyer, H., Reichert, C., Djajadihardja, Y., and Prexl, A. (2002). Tectonic features of the southern sumatra-western java forearc of indonesia. *Tectonics*, 21(5):15 PP.

Schoeffel, I. and Das, S. (1999). Fine details of the wadati-benioff zone under indonesia and its geodynamic implications. *Journal of Geophysical Research*, 104(13):13101 – 13114.

Scholz, C. and Small, C. (1997). The effect of seamount subduction on seismic coupling. *Geology*, 25:487–490.

Seno, T. (2002). Tsunami earthquakes as transient phenomena. *Geophysical Research Letters*, 29(10):4 PP.

Shulgin, A., Kopp, H., Mueller, C., Lueschen, E., Planert, L., Engels, M., Flueh, E., Krabbenhoeft, A., and Djajadihardja, Y. (2009). Sunda-banda arc transition: Incipient continent-island arc collision (northwest australia). *Geophysical Research Letters*, 36(L10304):6 PP.

Shulgin, A., Kopp, H., Mueller, C., Planert, L., Lueschen, E., Flueh, E., and Djajadihardja, Y. (2010). Structural architecture of oceanic plateau subduction offshore eastern java and the potential implications for geohazards. *Geophysical Journal International*.

Sieh, K. and Natawidijaja, D. (2000). Neotectonics of the sumatran fault. *Journal of Geophysical Research*, 105(12):28295–28326.

Stein, S. and Wysession, M. (2003). *An introduction to seismology, earthquakes, and earth structure*. Wiley-Blackwell.

Susilohadi, S., Gaedicke, C., and Erhardt, A. (2005). Neogene structures and sedimentation history along the sunda forearc basins off southwest sumatra and southwest java. *Marine Geology*, 219:133–154.

Taira, A., Byrne, T., and Ashi, J. (1992). *Photographic atlas of an accretionaryprism: geologic structures of the Shimanto Belt, Japan*. University of Tokyo Press.

Tanioka, Y. and Satake, K. (1996). Tsunami generation by horizontal displacement of ocean bottom. *Geophysical Research Letters*, 23(8):861–864.

Tapponnier, P., Peltzer, G., Le Dain, A. Y., Armijo, R., and Cobbold, P. (1982). Propagating extrusion tectonics in asia: new insights from simple experiments with plasticine. *Geology*, 10(12):611–616.

Taylor, F., Bevis, M., Schutz, B., Kuang, D., Recy, J., Calmant, S., Charley, D., Regnier, M., Perin, B., Jackson, M., and Reichenfeld, C. (1995). Geodetic measurements of convergence at the new hebrides island arc indicate arc fragmentation due to an impinging aseismic ridge. *Geology*, 23:1011–1014.

Taylor, F., Mann, P., Bevis, M., Edwars, R., Cheng, H., Cutler, K., Gray, S., Burr, G., Beck, J., Phillips, D., Cabioch, G., and Recy, J. (2005). Rapid forearc uplift and subsidence caused by impinging bathymetric features: Examples from the new hebrides and solomon arcs. *Tectonics*, 24:23 PP.

Toomey, D., Solomon, S., and Purdy, G. (1994). Tomographic imaging of the shallow crustal structure of the east pacific rice at 9°30'n. *Journal of Geophysical Research*, 99:24,135–24,157.

Tregoning, P., Brunner, F., Bock, Y., Puntodewo, S., McCaffrey, R., Genrich, J., Calais, E., Rais, J., and Subarya, C. (1994). First geodetic measurement of convergence across the java trench. *Geophysical Research Letters*, 21(19):2135–2138.

Tsuji, Y., Imamura, F., Matsutomi, H., Synolakis, C., Nanang, P., Jumadi, S., Harada, P., Han, S., Arai, K., and Cook, B. (1995). Field survey of the east java earthquake and tsunami of june 3, 1994. *Pure and Applied Geophysics*, 144(3-4):839–854.

Uchida, N., Kirby, S., Okada, T., Hino, R., and Hasegawa, A. (2010). Supraslab earthquake clusters above the subduction plate boundary offshore sanriku, northeastern japan: Seismogenesis in a graveyard of detached seamounts? *Journal of Geophysical Research*, 115:13 PP.

Van Avendonk, H. (1998). *An investigation of the crustal structure of the Clipperton transform*. PhD thesis, University of California, San Diego.

Van Avendonk, H., Harding, A., Orcutt, J., and Holbrook, W. (2001). Hybrid shortest path and ray bending method for traveltime and raypath calculations. *Geophysics*, 66:648Ü653.

van Hunen, J., van den Berg, A., and Vlaar, N. (2002). On the role of subducting oceanic plateaus in the development of shallow flat subduction. *Tectonophysics*, 352:317–333.

von Huene, R. (2008). When seamounts subduct. *Science*, 321(5893):1165 – 1166.

von Huene, R. and Lallemand, S. (1990). Tectonic erosion along the japan and peru convergent margin. *Geol. Soc. Am. Bull.*, 102:704–720.

von Huene, R., Ranero, C., and Scholl, D. (2009). *Convergent margin structure in high-quality geophysical images and current kinematic and dynamic models*. Springer Verlag.

von Huene, R., Ranero, C., and Vanucci, P. (2004). Generic model of subduction erosion. *Geology*, 32(10):913–916.

von Huene, R., Ranero, C., and Weinrebe, W. (2000). Quaternary convergent margin tectonics of costa rica, segmentation of the cocos plate, and central american volcanism. *Tectonics*, 19:314–334.

Wagner, D., Koulakov, I., Rabbel, W., Luehr, B., Wittwer, A., Kopp, H., Bohm, M., and Asch, G. (2007). Joint inversion of active and passive seismic data in central java. *Geophysical Journal International*, 170(2):923–932.

Walther, C. (2003). The crustal structure of the cocos ridge off costa rica. *Journal of Geophysical Research*, 108:21 PP.

Wang, K. and Hu, Y. (2006). Accretionary prisms in subduction earthquake cycles: The theory of dynamic coulomb wedge. *Journal of Geophysical Research*, 111:16 PP.

Weinzierl, W. (2010). *Crustal structure of the central Lesser Antilles island arc: seismic near-vertical and wide-angle profiling*. PhD thesis, IFM-GEOMAR, Kiel.

Werner, R., Hauff, F., and Hoernle, K. (2009). *SO-199 CHRISP: Christmas Island Seamount Province and the Investigator Ridge: age and causes of intraplate volcanism and geodynamic evolution of the south-eastern Indian ocean*. Number 25. IFM-GEOMAR.

White, R., McKenzie, D., and O' Nions, K. (1992). Oceanic crustal thickness from seismic measurements and rare earth element inversions. *Journal of Geophysical Research*, 97(13):19,683–19,715.

Zelt, C. (1999). Modelling strategies and model assessment for wide-angle seismic traveltime data. *Geophysical Journal International*, 139(1):183–204.

Zelt, C. and Barton, P. (1998). Three-dimensional seismic refraction tomography: A comparison of two methods applied to data from the faeroe basin. *Journal of Geophysical Research*, 103:7187–7210.

Zelt, C., Sain, K., Naumenko, J., and Sawyer, D. (2003). Assessment of crustal velocity models using seismic refraction and reflection tomography. *Geophysical Journal International*, 153(3):609–626.

Acknowledgements

I would like to thank Prof. Dr. Heidrun Kopp for her continuous support and her unrestless and patiently advice over the past years. She provided many new ideas in helpful discussions which often clarified important aspects of this work. This is very much appreciated. I thank Prof. Dr. Ernst Flueh for his suggestions and acting as my coreferee.

Thanks to all the other members of the MERAMEX project of the geophysics department at the University of Kiel Prof. Dr. Rabbel and Dr. Diana Wagner, the GFZ members Birger Lühr and Dr. Ivan Koulakov and Dr. Udo Barckhausen from the BGR. Many thanks to the participants of RV 'SONNE' Cruise 179, scientists and crew, for the acquisition and pre-processing of the data.

Many thanks to Dr. Martin Scherwarth for the numerous suggestions to improve the manuscript. I would like to thank Alexey Shulgin for supporting me with some figures and discussions during the final phase of my work. Thanks to all the members of the Marine Geodynamics Department for their assistance and the nice working atmosphere in recent years.

Last but not least, I would like to thank my wife Claudia for her patient support and her encouragement and active assistance especially in the turbulent phase of the last months – and even more for all the things beyond.

I want morebooks!

Buy your books fast and straightforward online - at one of world's fastest growing online book stores! Environmentally sound due to Print-on-Demand technologies.

Buy your books online at
www.morebooks.shop

Kaufen Sie Ihre Bücher schnell und unkompliziert online – auf einer der am schnellsten wachsenden Buchhandelsplattformen weltweit! Dank Print-On-Demand umwelt- und ressourcenschonend produziert.

Bücher schneller online kaufen
www.morebooks.shop

KS OmniScriptum Publishing
Brivibas gatve 197
LV-1039 Riga, Latvia
Telefax:+371 686 204 55

info@omniscriptum.com
www.omniscriptum.com

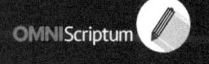

Printed by Books on Demand GmbH, Norderstedt / Germany